I Think and Write, Therefore You Are Confused

I Think and Write, Therefore You Are Confused

Technical Writing and The Language Interface

Vahid Paeez

CRC Press
Taylor & Francis Group
Boca Raton London New York

CRC Press is an imprint of the
Taylor & Francis Group, an **informa** business

First edition published 2022
by CRC Press
6000 Broken Sound Parkway NW, Suite 300, Boca Raton, FL 33487-2742

and by CRC Press
2 Park Square, Milton Park, Abingdon, Oxon, OX14 4RN

© 2022 Vahid Paeez

First edition published by CRC Press 2022

CRC Press is an imprint of Taylor & Francis Group, LLC

ISBN: 9780367754402 (hbk)
ISBN: 9781032048390 (pbk)
ISBN: 9781003194835 (ebk)

Typeset in Times
by Deanta Global Publishing Services, Chennai, India

Dedication

In Loving Memories of:

Dr. Richard E. Palmer (1933–2015)

and

Dr. Lewis W. Heniford (1928–2018)

who showed me where to look at but not what to see.

And for

Dr. Earl R. Anderson and Dr. Gerald L.
Bruns who aspire to inspire.

It is only by putting it into words that I make it whole; this wholeness means that it has lost its power to hurt me; it gives me, perhaps because by doing so I take away the pain, a great delight to put the severed parts together. Perhaps this is the strongest pleasure known to me. It is the rapture I get when in writing I seem to be discovering what belongs to what; making a scene come right; making a character come together. From this I reach what I might call a philosophy; at any rate it is a constant idea of mine; that behind the cotton wool is hidden a pattern; that we – I mean all human beings – are connected with this; that the whole world is a work of art; that we are parts of the work of art [...] We are the words; we are the music; we are the thing itself. And I see this when I have a shock.

(Virginia Woolf, *Moments of Being* 1985)

And then again, there can be said to be three kinds of author. Firstly, there are those who write without thinking. They write from memory, from reminiscence, or even directly from other people's books. This class is the most numerous. Secondly, there are those who think while writing. They think in order to write. Very common. Thirdly, there are those who have thought before they started writing. They write simply because they have thought. Rare. Even among the small number of writers who actually think seriously before they start writing, there are extremely few who think about the subject itself.

(Arthur Schopenhauer, *Essays and Aphorisms* 2004)

I'm sitting here writing. (Language.) I'm writing this newspaper column and will receive money for it. (Work.) A column about my philosophy of life. (Feelings.) I (I), who once was born and immediately found myself among my fellow creatures (Humans), am sitting here writing in a Copenhagen apartment. Someone must have built it at some time (Social System). It has a history. A life span, one of many on the earth (Nature). Earth, which we can now see from space. There it hangs. That's where I live (Universe).

(Inger Christensen, *The Condition of Secrecy* 2018)

Thus is revealed the total existence of writing: a text is made of multiple writings, drawn from many cultures and entering into mutual relations of dialogue, parody, contestation, but there is one place where this multiplicity is focused and that place is the reader, not, as was hitherto said, the author. The reader is the space on which all the quotations that make up a writing are inscribed without any of them being lost; a text's unity lies not in its origin but in its destination. Yet this destination cannot any longer be personal: the reader is without history, biography, psychology; he is simply that someone who holds together in a single field all the traces by which the written text is constituted.

Which is why it is derisory to condemn the new writing in the name of" a humanism hypocritically turned champion of the reader's rights. Classic criticism has never paid any attention to the reader; for it, the writer is the only person in literature ... We know that to give writing its future, it is necessary to overthrow the myth: the birth of the reader must be at the cost of the death of the Author.

(Roland Barthes, *Image-Music-Text* 1977)

The fish trap exists because of the fish; once you've gotten the fish, you can forget the trap. The rabbit snare exists because of the rabbit; once you've gotten the rabbit, you can forget the snare. Words exist because of meaning; once you've gotten the meaning, you can forget the words. Where can I find a man who has forgotten words so I can have a word with him?

(Zhuangzi, *Basic Writings* 2003)

Contents

PART I   I Write, Therefore I Am Misunderstood

PART II   Language, Would You Mind Stop Speaking for Me?

A Preface to Confess

Let me tell you a story. 52 Blue is the world's loneliest whale, a blue whale that sings at 52 hz, which is too high to be heard by other blue whales. This whale was recorded in 1992 and was introduced to the world in 2004. On social media, pages have been set up for this solo traveler. But why is he so lonely? Doesn't he have the appearance of a whale like other whales? Doesn't he wish to connect with his fellow whales? Well, he does. But communication that goes unheard will also end up being unknown, unidentified, unnoticed, and one way. One-way communications are doomed in any collaborative context in which there is a one plus another available. Although you do have a voice, you'd end up being lonely if you fail to make your voice heard, to get the message across, to allow language and words to deviate from your thoughts and intentions. If you choose not to be on the same wavelength as your listener or reader, then you'll have no audience. In communication, writers that deviate from the expected, standard, and known frequencies will end up being solo travelers. Your reputation gets damaged and, on top of all, many other people continue to suffer by losing confidence in your writing and finding an alternate way to close the gap you were under the impression of closing.

I must confess to a love of human factors and writing. Writing brought me into the United States. It was the right message to the right people at the right time with the right intention translated into the right words. Yes, I did have a voice and I made it heard. We all have voices that could be heard or could get lost. Adam spoke, so did Noah. We learn from the scriptures that Adam was directly taught by God himself. When God formed all living things, according to the Bible, He then assigned the naming of them to Adam.

Language signs you up and drops you out. You could become an insider and an outsider only by the language you know and the language that knows you. Language does not convey meaning in an objective sense. Meaning – produced by the sender and reproduced by the receiver – precedes language. Language aims at translating thoughts and intentions into words – a process in which meaning could be lost. Language is more than just a means of communicating. It is one of the most important and revealing social manifestations of who we are and where we come from. To criticize someone's language assumes a position of superiority and a right to judge that person. It is never just about language. Rather, marginalization through language becomes a proxy for racism, homophobia, xenophobia, and elitism.

The coronavirus pandemic is obliging us to wear facemasks and maintain a distance of 6 ft from others, which is also drawing attention to the invisible masks we have been wearing for as long as we are aware of – masks of miscommunication, misinformation, misinterpretation, misjudgment, and misunderstanding – and to distances we have deliberately or inadvertently been maintaining from one another to communicate effectively or to communicate at all.

Confined in the prison house of language, once you set pen to paper or use a keyboard or a keypad to type, the act of writing, preceded by the act of thinking, takes place. But when you're done, you may still own the pen, the paper, the keyboard, or the keypad, but NOT the words. They belong to the powerful system of reading and interpretation of which misreading and misinterpretation are also inseparable components. In the global village we're living in – especially during this COVID-19 pandemic – writing does play an even more important role than before. With the growing number of people being put on telework, email has become an essential part of day-to-day conversations with our coworkers, supervisors, customers, and all stakeholders. Clarity in correspondence should not be underestimated. When I was going to English school (as a foreign language), I was taught that English is not the language of unbreakable rules. Later on, life taught me that in order to break a rule, you must first master it. Whether you have a role or a say in creating the Before the Fact documents (requirements, etc.) or in the After the Fact documents (technical manuals, documents, etc.), this book is for you.

My background is primarily in writing (documenting, creating, editing, revising, disseminating, and training of the requirements) and the human factors studies. The first standard operating procedure that I created was *SOP 0.1: An SOP on How to Write an SOP*. Maintaining that we humans are either readers or writers or both, I should say that we all do have one thing in common: we are translators. We are constantly reading and judging our surroundings, people, and events and translating them in order to perceive, understand, interpret, make sense, and seek patterns by imposing order on chaos. Being a meaning-maker reluctant to put up with things that do not conform to its principles and rules of acceptance, our brain does continuously impose order upon what it interprets as chaos. Writers can be the worst culprits in creating such chaos.

We communicate, miscommunicate, interpret, and misinterpret. We are not alone when writing or when reading. Our past experience, expectations, biases, misconceptions, and misjudgments are hard and, silently, at work. Language is not neutral. It reflects cultural values and influences how we see the world. Language is everywhere and it stalks us. It has surrounded us, captured and owned us entirely. Being deceivingly flexible, language also gives way to misunderstanding. Your art rests in the *how* and not in the *what* of this beautiful challenge. When we are writing, it's just us and the words. When our audience is reading our writing, it's them and the words. The atmosphere that we create to breathe the linguistic air must be identical to the atmosphere our reader (or user) will also require in order to survive.

The inspiration for this book came after the numerous professional experiences I have had in the United States, as an immigrant, and in the Middle East. Communication is more than just a way to get ideas across or exchange points of view. It is the process by which we interact with others and seek or give information essential to our daily lives, allowing us to control the circumstances in which we function. If communication is the transfer of meaning, then for successful communication to occur, you should understand something exactly as I do, for the experience to be shared and for the message to be fully conveyed, acknowledged, and complied with.

Like many of you, I have had people working in my house, most recently a plumber and his assistant to replace the water heater as well as taking care of a couple of other things. It took about 30 minutes to replace the cartridge in my shower, which would have taken me, at least, 5 hours. Why? He has been doing it for many years. If I had tried to do it myself and save some money on the labor, I had to refer to either the written instructions/user-guide – which are not always easy to follow – or google it, watch a few videos on YouTube to familiarize myself with the steps, tools, etc., first – all of which could have never guaranteed my performing the task successfully without making a mess somewhere else. Being a process-driven guy and putting my quality assurance (QA) hat on, I asked Mr. Plumber about the steps written in the user-guide. His response was, "Well, I have never read those. I learned from my father how to do it." Although this person can communicate well verbally, writing, for him, is a serious challenge. He was not even happy after he went over a few words from the user-instructions on the box. I have asked other handymen who worked in my house to give me written instructions for what they did or were doing, but not one was able to provide any. I posed in front of them with my fancy pen and notebook, but I still failed. The electrician said, "I know how to get the job done, but I cannot teach another person or write down the steps for you." Not that he didn't want to, he was actually unable to. Writing can cripple our mind – both the writer's and the reader's. I have had the same issue with my coworkers, supervisors, the audit/inspection teams, and with my students.

I do have another intriguing anecdote to share. Aren't we storytellers?

I vividly recall that in one of my jobs in the United States, I met a colleague whose command of English impressed me tremendously. I, an avid reader and lover of languages, couldn't help writing down the (new) words, phrases, and idioms he used during the conversations he had with us in person or over the phone with other people. Every day, he did have a new lesson, admired and cherished by me, without even knowing about it. I couldn't wait for him to talk so I could listen to the beauties of the English language and add more vocabulary and phrases to my glossary of technical terms and jargon. Having taught English classes for different age groups and positions for a long time, and having delved into technical publications after becoming a technical writer in the aviation and aerospace industry later on, I made a confession to myself, concluding that being a native English speaker must have something to do with such an expertise. I continued to listen to John, the well-spoken coworker, and enjoyed it until the day of disillusionment. I am alluding to that Tuesday morning when I opened my email and saw the first email from him – I had been CC'd on an email to three other persons. I had never been so shockingly dumbfounded before. Being a big advocate of writing in English and having graded a ton of student papers in the past, I saw John's email to be nowhere even close to average writing. Basic rules of grammar had been overlooked, both semantically and syntactically, and his message was in violation of conventions of standard writing. It conveyed nothing but ambiguity, both in form and content. But why? Wouldn't writing be as easy and fluid as speaking? No. No. No. He was a perfect verbal communicator, but not competent in writing. That's a common issue these days. Speaking is naturally acquired. Writing is not.

There are good examples of bad writing – we've come across them and will continue to run into them at home, at work, in the society, on the bus, on the train, at the airport, on the plane, in the United States, and in other countries. Like many, I have utilized hundreds of examples in my personal and professional life. I remember once I purchased a pocket solar calculator to perform some aircraft weight and balance calculations. The lengthy printed instructions inside the box never helped me think outside the box. They would only add to my confusion because they had been very likely not meant otherwise. Was a printed guide included in the box to walk me, the user, through steps? Yes. Hadn't a team of writers created that? Most probably. Did the user (me) find it useful? Of course NOT. Had they created that user-guide for me? Absolutely NOT.

Part of the training that I have developed and conducted has circled around the user-centered writing that gets immediately tested on the spot through objective validation by other users. I want to show you ways and examples of not only different kinds of writing, but also how technical publications are laid out in the civilian and also the non-civilian sectors in the aerospace industry. The regulatory agencies are facing similar challenges. I have examples of how they are violating their own prescribed requirements. How does effective communication take place? Many individuals and teams write, edit, and proofread, but would all this also enable them to communicate? Is technical writing being taken as seriously as the new advancements in technology are? Whether handwritten, electronically mediated, or face-to-face, communication is more than a value-neutral exercise in information transfer; it is a complex social transaction.

In writing, with the nonverbal aspect of communication (using hands and face to express or clarify meaning) being automatically omitted, readers and writers will need to learn how to properly use this powerful tool. To pursue this, they should fully familiarize themselves with the tool in a crawl, walk, run fashion. I compare good writing to a game of soccer. A soccer team is clearly an interdependent team, i.e. no player, no matter how talented, has ever won a game by playing alone. Unlike weightlifting, in soccer, no significant task can be accomplished without the help of all team members who typically specialize in different tasks (goal keeper, forwards, defenders, etc.). Writing in a vacuum is the replacement of soccer with weightlifting. It's like celebrating independence where interdependence is actually desired.

Let's explore this. Language is like a toolbox of labels and tags we carry and constantly affix to things and events. It is the most important medium for relating to the world. For me as an immigrant who speaks English as a second language, the English language is an international commodity that can be obtained freely by individuals, groups, teams, and nations. This precious commodity is not being properly treated by the English-speaking countries anymore. One of the delights of English is its rich vocabulary, which provides a selection of words with which we're given the opportunity to express our thoughts. We use language, on a day-to-day basis, to translate our thoughts and share them with others through this powerful medium.

There are plenty of signs everywhere for someone to read and pay attention to, yet I very rarely see it happen. If drivers paid enough attention to all traffic signs – some very easy to miss – a much better teamwork would have formed among drivers.

Useless requirements are a check in the box for cover your ass (CYA) purposes against liability, while creating requirements to share a useful message with others in a useful manner. The last time I went hiking, I ran into some new signs as a result of the COVID-19 pandemic. Although the quantity of signs would be intimidating to the human mind, and counterproductive in paying quality attention to them, I still read them and, of course, think of better versions for them. The other day, I was hiking in a new place with my wife and ran into a safety sign, very critical (Figure 0.1).

Besides missing some good pictures to illustrate some of the bullets in the sign below, I immediately asked my wife about the possible ways of negotiating your life with a hungry mountain lion when I read the fifth bullet.

When I first learned about this clause, "Tech pubs suck," I was astounded because I heard it from those individuals in charge of training people and responsible for encouraging them to use those tech pubs. When I heard this statement also from other technical people and teams, I drafted an article in my mind that revolved around themes such as: you don't write with your hands, you write with your mind,

FIGURE 0.1 Caution – a sign to read, understand, and comply with

with your expectations, with other people's expectations, with your hidden, unde-sired biases, with your engagement with the topic, with your detachment from the concept, with subjective and objective viewpoints, with your illusion of communica-tion that would communicate misinformation.

In order to solve a problem, first you must know that there is a problem. Merely talking about problems won't solve them. My goal is to prevent this from happening to you by showing you where the trouble rests. Language could be a lens through which could shine the light of reason. It oils social wheels. Tech Pubs Suck implies that "I do know of a better version." But that "better version" could still be tagged as a bad (if not worse) version in the eyes of other individuals or groups of individuals.

Acknowledgments

Working on this book, I have had the good fortune to benefit from the assistance of friends, colleagues, students, and those who taught me something without having the intention to. I would like to thank all of the contributors to this book for their professionalism, insight, diligence, words, silence, action, and inaction.

This book could not have been written without the support of a host of experts who have generously passed along their own knowledge that has found its way into the pages that follow: Rob Manning, Tim Baker, John Doherty, Dr. Paul Lawley, Dr. Gerald L. Bruns, Dr. Karen Rose, Dr. Margaret Magnus, Dr. Tim Morris, Weed, James Blaylock, Dara Sabahi, James Evans, Peter Muse, Keith Glassman, Bo Tubbs, Dr. Jalal Sokhanvar, and Dr. Amir Ali Nojoumian. To those professionals I extend my heartfelt thanks. To Language, I surrender and ask for blessings.

I also thank my wife Shadi for her encouragement and support.

Vahid Paeez
January 2021

Acknowledgements

Author Bio

My name is Vahid Paeez. I am from the land of the Thousand-and-One-Nights, having my feet on the third planet from the sun, with the precious gift of a curious mind. I must confess to a love of language and writing, human factors studies, effective communication, literature and philosophy, air and space. I have authored, co-authored, and translated articles and chapters in technical and non-technical areas, from human–machine interface, aircraft system safety, and continued airworthiness to literature and fiction. My background is primarily in writing (technical, non-technical, creative), quality assurance, training, research, and development.

As a member of the ASQ (American Society for Quality), I was selected as one of their New Voices for Quality (a Fresh Face) in 2016. Being a member of the AIAA (American Institute of Aeronautics and Astronautics), I have led the human factors sub-technical committee for this nonprofit organization. Also, I received the AIAA's 2018 Award in San Diego for *Outstanding Contribution to the Community*. Working as a quality assurance specialist, an instructor, a human factors researcher and, of course, a technical writer (in two continents, two countries, and many cities), I have had the opportunity to work with various teams, supervisors, individuals, and groups – from different cultures and backgrounds – who had to communicate in English, whether in the form of writing or speaking.

To me, we humans have one thing in common: we are inexhaustible translators. Along with our biases and misconceptions, we are constantly reading, perceiving, and judging our surroundings to make sense of events and to understand patterns. Writing experimental fiction, and technical documents in two languages, I find myself to be a good example of that missing link between fiction and nonfiction. I take pride in my achievements and in having a strong sense of purpose. I strive to do for my mentees the exact thing my mentors have done for me: show them where to look at, but not what to see.

Part I

I Write, Therefore I Am Misunderstood

1 Requirements
Meant for All Seasons

1.1 HOW OLD IS THE OLD HOW?

Noah must have managed one of the earliest recorded projects in the Bible – *building of the Ark*. Intent on a great project to renew the earth and the Adamic covenant, God calls upon Noah, the one man He can trust, to carry out His plans. However, God does not just tell Noah, "Now go and build an ark." God details what He wants done. In the first phase of the project, He goes into a lot of detail and He also explains to Noah why each element is required to get the job done. Having explained the basic requirements of the Ark, God moves on to the main processes of phase two of the project, beginning with His own action items. He then returns to what He requires of Noah (gather two of every animal on earth; all the necessary food and water).

> Make thee an ark of gopher wood; rooms shalt thou make in the ark, and shalt pitch it within and without with pitch. And this is the fashion which thou shalt make it of: The length of the ark shall be three hundred cubits, the breadth of it fifty cubits, and the height of it thirty cubits. (Genesis 6:14–15 King James Bible 2020)

Because God crafts a quite compelling narrative as an effective communicator, Noah clearly understands what he is supposed to do and, of course, why. The project is a success: "all that God commanded him, so did he" (Genesis 6:22) (Figure 1.1).

In my years of working as a QA auditor and configuration management specialist, I have learned that undocumented requirements are rumors. Poorly documented requirements are confusion. Any form of linguistic creation that fails at the receiving end due to any reason is frustration.

1.2 THE MIRROR AND THE LAMP

Being a unique cognitive ability, language connects us by creating a context that disconnects us from the rest. Language does discriminate. It is a distinguishing attribute of our species (apes cannot possibly acquire it). Language brings order to chaos. It brings peace and calm, when designed so. Language relieves and hurts. Language revives and murders. It is a vital component of human autobiographical memory. We humans develop and deliver countless designs or simply use them during our lifetime. Not only is language specific to the human race, but all human beings have language. Communicating by means of language is the most characteristically human of all man's behavior patterns.

FIGURE 1.1   An Egyptian inspector is supervising the making of building blocks intended for a pyramid. The man to the left and the man on top of the block are working on the stone. The man to the right is measuring the block with a rope. This illustration was found in a king's grave in Thebes around 1450 BC.

You mirror an existing light when speaking – it's a common game played by many, and masterfully played by John, the coworker I had. It's the art of the mass. But to me, writing resembles that lonely, individual act that starts after lots of other things have taken place. It is preceded by steps. Documents are born, yet they need to be properly conceived. Premature birth would have complications. The act of writing is the last thing in the process of writing – exactly similar to the numerous activities and events taking place behind the scenes before the curtain goes up for the performance to begin. Your strength in writing would depend upon planning, design, and development, rather than the delivery phase. In speaking, your delivery will be accepted, although if not 100% satisfactorily, as there's quite a brief amount of time between the development and delivery phases. Besides, taking advantage of nonverbal skills in the very context, along with clarification through rephrasing and asking and answering questions for more clarity, is granted as an available option to the involved parties and participants.

If you're choosing to communicate through language, then use acceptable language to communicate. You do have the right to write, but do it right. Declaration of independence by the text you have produced would be a milestone to accomplish. Shoot carefully for the survival of your text, without you. Writers perish when their writing is delivered. The more independently your reader (user) can survive and do without you, the better of a mission you have accomplished. In his essay, "The Death of the Author," Roland Barthes concludes, "The birth of the reader must be at the cost of the death of the Author" (Barthes 1977). Direct your reader (user) toward convergence. Avoid divergence.

Singing and writing are more difficult and challenging than speaking.

1.3 SIMPLIFIERS VS. COMPLEXIFIERS

It goes without saying that technology has transformed the way we communicate. We send and receive more and more emails every day. We text. We tweet. We have reduced our communications down to efficient sound bites. In the technical field, concepts continue to become more and more complicated. However, your art as a writer would rest upon keeping the syntax – your vehicle – as simple as possible. Regardless of how sophisticated the concept we're trying to write about, our mission must focus on writing as humans for humans in the digital age.

We think through interpretative filters that shape our perception of reality. Good writers facilitate smooth conversion of perception into reality. Otherwise, this could be tricky and powerful enough to change an objective experience into a subjective experience. As unexplainable as it may sound, never give up. All you need to do is naturalize the supernatural and make the unfamiliar familiar. Make the complex simple. Using a tech pub is no different from using a vending machine – they both exist to obviate some need. Writing could be a habit, but writing right always remains a choice.

Language can become a barrier to logic when failed to be used properly, purposefully, and adequately. Miscommunication would not hesitate to emerge if words stop creating and start to miscreate.

1.4 THE ORIGINAL SCENE: LANGUAGE SPEAKS

The Old Testament shows us what follows when the authority is disobeyed. It is in Genesis (2:16–17) when God speaks to Man for the first time – we are not told in what language – and warns him not to eat of a (certain) tree that would lead him to a certain and inevitable death. However, why Adam is not told about Satan who seems to be fully aware of God's creation in detail in general is not clear. When the voiced requirements are inadequate on the one hand and you would love to have your eyes opened and become expert in knowing good and evil on the other hand, you would end up listening to other voices. Falling from grace is the price you pay, as did Adam and Eve, by listening to another voice (Satan) and eating the Forbidden Fruit.

Creation, according to the Old Testament, is preceded by an act of speech, "Let there be light" (Genesis 1:3). In his "Letter on Humanism," Martin Heidegger (1889–1976) writes, "Language is the house of Being. In its home man dwells. Those who think and those who create with words are the guardians of this home" (Heidegger 1976). Being the most central characteristic of the human species, language is a quick and painless way of passing on the discoveries of one generation to the next.

Language is the basis of social life and living together. Thanks to language, our internal world becomes visible, audible, and tangible to others. Language's mission is to make understanding and being understood possible, despite implications.

I am sure that you, like me, have come across bizarre words and strange phrases (how they are categorized as "bizarre" and "strange" being a thing of perception) in different kinds of writing, depending on their contexts. In technical pubs, this would be unacceptable, though. It does keep occurring, however. Although Microsoft Word does have a built-in spelling and grammar checker that can be very helpful for both writers and editors to review their document prior to finalizing it, misspellings – e.g. hanger (instead of "hangar"), brake time (instead of "break time"), and it's (contraction of "it is" or "it has" and not equal to "its" as a possessive determiner) – as well as unclear, vague requirements still deem unavoidable and would go unnoticed if you relied too much on the machine before you. This has turned into a recurring phenomenon, regardless of how prohibited committing such mistakes continues to remain in force.

Although misunderstanding often results from misusing language, you cannot help blaming the commander named language and his crew (words). Language can be shared with others in different forms, but meaning is not found, it is made.

2 The Language of Language

People communicate for a number of reasons. Sometimes they simply want to share their perspective. More frequently, though, speakers or writers want their listener or their audience to respond by doing something (taking action), such as following directions, fulfilling expectations, or providing assistance. Understanding requires team efforts. As a writer or speaker, you do become a significant part of the interpretation process to create meaning if you devise language properly. Rumi says, "Words can't contain the depths of meaning's sea; Expression can't reach meaning's depth."

Language development moves from simplicity to complexity in humans. We sample tens of thousands of words in our memory, combine them with near-perfect syntax, and spew them out because not having a voice would equal non-being.

Meaning precedes language. Language is only the tip of a spectacular cognitive iceberg, and when we engage in any language activity, be it technical or artistically creative like a poet or an experimental writer, we draw unconsciously on vast cognitive resources. We are owned by the miraculous power of words and language to the extent that even having inner speeches and thoughts would be impossible without language (Figure 2.1).

An English friend of mine says he spent much time, when younger, thinking about how differently the rich and poor people spoke. Where he was brought up, most people spoke a regional dialect, and written English was almost a foreign language – they would be punished at school for their "mistakes," for not writing "proper" English, so naturally they had little love for it. Despite the genius of wordsmiths like William Shakespeare, England's bilingual nature (which continues to this day) can be traced back to the Norman Conquest of AD 1066, as per Weed, my English friend. The 20th century rolled on, with its radio and films and television, and just as it seemed a more homogenous form of English might arise, along came waves of immigrants, and within a few generations, a new urban patois had developed, a mixture of working-class English, West Indian, African American, and Asian. According to Weed, at some point their education system gave up and began recognizing it as a valid form of communication, especially in plays and poetry and, of course, song lyrics.

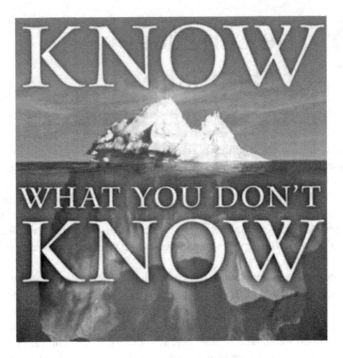

FIGURE 2.1   The language system: the known vs. the unknown.

2.1   WORDS WORDS, FRIENDLY FOES

Before you speak, let your words pass through three gates:

> At the first gate, ask yourself, "Is it true?"
> At the second gate, ask, "Is it necessary?"
> At the third gate, ask, "Is it kind?"

– Sufi saying

What is language? Well, I should say that language is the same phenomenon that allows you to ask such questions. The ironic part is that you would also have to use language to provide answers to such questions. Weird, huh? Where do words begin and end? Being a sequence of letters or sounds with meaning, it seems perfectly obvious that pencil is a word, licnep is not; pencil has a meaning, licnep does not. Words can generally be broken down into units, called morphemes – the smallest meaningful chunks into which you can divide a word.

Language doesn't talk about reality; it is the reality. We read, misread, interpret, misinterpret, communicate, miscommunicate, but we refrain from being silent because our existence gets defined by words, is measured in words, with words, and through words. Meaning, again, is not found; it is made in a collaborative act between the sender (writer and speaker) and the receiver (listener, reader, observer).

Friedrich Nietzsche once said, "There are no such things as facts, only interpretations" (Nietzsche 2003). That's who we are and have always been: translators and interpreters. We constantly perceive (read) the world and interpret (translate) events, in a biased way. There are more than 6,000 languages in the world that are structurally diverse. English language, being an important member of the big family of languages, has many words with ambiguous meanings. For instance, peck can mean one-fourth of a bushel, or to strike with a beak. Rose can mean a flower or the past tense of the verb rise. Desert can mean a dry stretch of land or (with a different accent) to abandon someone. Brake and break cannot be used interchangeably. Besides, most speakers of English would not pronounce the "l" in the word "should," nor would they pronounce the "w" sound in the word "written." Clearly, then, the English language is not pronounced as it is written.

"Get" has many meanings and usages in English. In context, however, listeners and readers usually understand the meaning. Being the ultimate goal of writing or speaking, meaning is highly desired, yet it can be brutally avoided.

Another word in the English language that could be a noun, verb, adjective, adverb, and preposition is "up." Look up the word "up" in a dictionary. It is said that this two-letter word in English has more meanings than any other two-letter word.

We seem to be pretty mixed UP about "UP."

If you are *up* to it, try building *up* a list of the many ways *up* is used. Well, it will take *up* a lot of your time, but it's *up* to you not to give *up*. You may end *up* having fun with looking up *up*. Your time is not *up* yet, so keep trying and continue to look *up*.

Dr. Samuel Johnson is said to have written the first proper dictionary in English – Dictionary of the English Language – published in London in 1755. Some of its definitions of English words show the changes of meaning that words have had since then.

2.2 CAPABLE TRANSLATOR CALLED LANGUAGE

2.2.1 ASSETS CALLED NONNATIVES

Language is a translation of mental activity into a fundamentally different medium. Besides words, mental activity consists of more elements – sensations, emotions, memories, images, moods, preconceptions, biases, and more. There is a widespread myth claiming that one primitive language existed until some event destroyed that unity. The Bible (Genesis 11:9) tells us that humans tried to build a tower which would reach the sky. God, in anger at this presumptuousness, "did there confound the language of all the earth: and from thence did the LORD scatter them abroad upon the face of all the earth."

Ever since the Tower of Babel perished, the Tunnel of Translation has cherished.

English is the language used by most scientists and engineers for international communication; but whether people are using English as their first or second language, they are most likely to understand plain words in simply constructed sentences. I have taken creative writing courses in the United States, with native professors as a nonnative voice among native classmates. Some of my classmates admired my work and found it poetic, some didn't. The professor, however, would tell me that there was a certain beauty in my work that made it poetic and unique. Well, how I see this is the inclusion of context, mental baggage, personal history, metaphors, as well as cultural history that we human beings carry with us all the time. Once those traces – linguistic or nonlinguistic – touch us, they will never fade away and will continue to shadow us, our deliveries and productions, in as many forms as imaginable – writing being a significant example. America has provided me with the opportunity to practice listening to and speaking in a couple of voices – first language and second language – to produce a third voice for which this book would be a good example. Yes, an immigrant in the United States of America continues to deal with three voices on a daily basis – voices that give birth to new voices.

The second-language users should use language effectively, but they don't have to try to be like native speakers of that language. Expressing your own identity through the second language would be totally fine. Combination of voices, if not turned into a conflict of voices, does create a new voice that is beautiful.

James Joyce is considered by many to be the greatest writer in the 20th century. In a letter, Joyce writes that "Writing in English is the most ingenious torture ever devised for sins committed in previous lives" (Beja 1992).

In his essay, "The African Writer and The English Language" (1975), Chinua Achebe writes,

> My answer to the question *Can an African ever learn English well enough to be able to use it effectively in creative writing?* is certainly yes. If on the other hand you ask: *Can he ever learn to use it like a native speaker?* I should say, I hope not. It is neither necessary nor desirable for him to be able to do so. The price a world language must be prepared to pay is submission to many different kinds of use.
>
> **(Achebe 2013)**

Achebe concludes his essay by unveiling the miraculous power and capacity in the English language:

> I feel that the English language will be able to carry the weight of my African experience. But it will have to be a new English, still in full communion with its ancestral home but altered to suit its new African surroundings.

Immigrants keep adding up layers to the depth of the target language.

2.3   THE POWER OF LANGUAGE: THE LANGUAGE OF POWER

Does language have power? Are words powerful? Well, your responses would be powerless if they were both negative. Language has power and power has language.

As a writer, you must be completely aware of such a power. Language is an empire. When you're following the cooking instructions carefully, or when you're carefully reviewing the steps in a standard operating procedure to troubleshoot some fault on the flight deck, you're being mesmerized by the miraculous power of language that drags you with it. Politicians' speeches go beyond this in being powerful. True! Language consoles, heals, wounds, and kills.

A COOKING SOP

How do you make a procedure repeatable? Do exactly as cooks do. A cooking recipe is a good example of a repeatable procedure.

Recipe for Cabbage Soup

1. Buy a big head of cabbage.
2. Cut the cabbage in half.
3. Cut the halves in half (now you have quadrants of a sphere).
4. Add 2 quarts of water and 3 tablespoons of salt to a large pot and boil the quarter heads for 15 minutes.
5. Let the pot cool to hand warm, and then mash the cabbage until it is disbursed in water so that the quarters are no longer recognizable.
6. Add these to the pot:
   - Olive oil: 3 tablespoons
   - Vinegar: 4 tablespoons
   - Garlic salt: 1 tablespoon
   - Paprika: 1 tablespoon
   - Beef bouillon cube: 1
7. Reheat the pot to 150–170°F for 2 hours.
8. Serve in shallow bowls (serves 2 people) and decorate it with sprigs of fresh parsley.

Harold Pinter's play *Mountain Language* is composed of four brief scenes set in a political prison in the capital city of an unnamed country that is under dictatorial siege.

Waiting to see her imprisoned son, an old mountain woman has been harassed by guards and their dogs, one of which has savaged her hand. Finally allowed to visit her son across a table with a guard standing over them, she is not allowed to speak to him and is silenced by the guard because her mountain language has been officially banned by the authorities. In vain, the son tries to reason with the guard who monitors their encounter. When the son claims a common humanity with the guard by remarking that he too has a family (when the guard refers to his "wife and three kids"), the soldier promptly reports this to his superior ("I thought I should report, Sergeant ... I think I've got a joker in here"). They both mention "family," but for one it's a sacred thing, for the other a joke.

2.4   LANGUAGE CHANGES WITHOUT CHANGING

Writing can be impactful. Words console. Words hurt. Words lead. Words mislead. Words promise. Words fulfill. Words clarify. Words complicate. Language begins. Language ends. Language creates. Language destroys. Words make the invisible visible, and the visible invisible.

There is a story about Franz Kafka, one of the most famous writers of the early 20th century, told by Dora Diamont, Kafka's lover, about a walk in a Berlin park in the last year of Kafka's life. The writer meets a little girl who is crying and sobbing desperately because she has lost her doll. Kafka comforts her by reassuring her that the doll is just traveling and has even sent him a letter.

"Do you have it with you?" Asks the little girl.

"No, I forgot it at home, but I'll bring it tomorrow." Responds Kafka.

Kafka goes home and composes the letter. On the following day, he finds the little girl in the park and reads the fictional letter to her. In Kafka's made up narrative, the doll explains that she had grown tired of living in the same family and needed a change of air, town, and country. Although she still thought very fondly of the little girl, she needed to be away from her for a time. She promised to write every day, reassuring the girl of her love. So Kafka wrote each day, describing countries, adventures, school, and people that he knew only in fantasy. The little girl soon forgot about the painful reality of her loss and thought only about the pleasure of listening to Kafka's fiction. How a piece of writing can lift your spirit. This game went on for at least 3 weeks. Kafka, aware of the fact that the ending must restore order in the girl's chaotic life, ultimately decided to have the doll marry. In her last letter to the girl, Mr. Messenger brought this message from the doll, "You will understand that in the future we must give up seeing each other" (Zilcosky 2003).

See how writing can change things dramatically?

2.5   NEURO-LINGUISTIC PROGRAMMING (NLP)

Without the top management buying into change, compliance, and promoting the culture, organizations would be very unlikely to flourish. To sell your brilliant ideas to the upper management in your chain, your most important medium would be your non-technical skills to communicate effectively and get exactly – or close – to what you want out of those management review meetings, sessions, discussions, etc.

As we discussed earlier, language can be used both to clarify and to obscure truth. Poor communication can cause serious blockages to progress in an organization. Our brain is always analyzing what's going on around us. It's trying to find similar things from our past and trying to line them up with each other. The subconscious mind has stored millions of conversations with other human beings, in different contexts. These conversations have become so routine that the mind has virtually fallen asleep and given up. Neuro-Linguistic Programming (NLP) was invented by two academics in the early 1970s. It was the brainchild of John Grinder, a professor and lecturer in linguistics, and Richard Bandler, a math student at the University of California with a strong interest in computer science and psychology. They then went on to study Virginia Satir, who developed conjoined family therapy, and Milton Erickson, who is the father of modern clinical hypnotherapy. The "Neuro" in the name refers to the use of our senses in assessing people and things around us. The "Linguistic" self-evidently relates to use of language, and "Programming" is about the adjustments we make in order to succeed in our goals.

For example, as a writer, your ultimate goal is to sell your commodity – a piece of tech writing – to your reader or user. That product should trigger mental, emotional, and physical responses in your audience. As a salesperson, you must pay attention to your buyer's behavior and mirror their style in receiving, acknowledging, and complying with your message. Always write with the reader in mind. You are writing for humans who will read, understand, and interpret with their emotions, biases, expectations, preexisting values, belief system, and their brain – a complicated meaning-making machine.

2.6 HYPNOTIZED BY LANGUAGE

Hypnotherapy is a powerful tool in observing and resolving hidden psychological issues in people needing help. Both conscious and unconscious states of mind can be altered through learning. Understanding the role and effect of language enables NLP to be applied as probably the most powerful tool you can use to achieve the outcomes you want in your working life. According to the NLP, in order to increase the likelihood of success, you should change your thoughts (intention) and behavior (action). The role of language would be undeniable in either. A good example can be seen in sport when, in order to win a match, different strategies are tried until the most effective is identified and then vigorously applied. Writing or speaking is part of a big match. Make your desired meaning stand out.

We are all aware of the effects of hypnotism and the conditioning process implied in it. We can build deep self-acceptance with hypnotic scripts. For example, by describing the problem in the past tense, the therapist creates the feeling that the problem is receding away. There is nothing mysterious about hypnosis; it is just a state of focused concentration. Being an excellent tool, hypnosis can help a patient remember long-forgotten incidents.

Verbalizing statements is the foundation for self-hypnosis or meditation, "I am going deeper and deeper." Or when you're being hypnotized by a hypnotist, you're exposed fully to the power of language and words: "Now, I will count backwards from one to three, and when I do, you will go into trance. One: close your eyes. Two: relax further. Three: drift deep into trance."

Desired mental state can be inculcated under hypnosis and positive affirmations can help you achieve your goals. An affirmation is a pithy statement of your outcome that assumes it is possible and achievable and keeps your mind focused on it. Again, what are these affirmations made of? Right. Made of words. This is a great manifestation of the power of words. Language has healed many wounded souls.

3 Design of Language; Language of Design

Human language is a system for expressing or communicating thought. We constantly, ceaselessly, tirelessly, and continuously attempt to produce meaning in a sign-system called language. We strive to be heard.

In 1916, Swiss linguist Ferdinand de Saussure (1857–1913) saw language as a system consisting of arbitrary signs, in which there is no direct link between the sign and the concept it evokes. Signs are phenomena that represent other phenomena. Absence of those phenomena has made the presence of signs a mandatory rule for humans to be able to communicate. Saussure called the structure of language *langue* (the French word for language), and he called the individual utterances that occur when we speak *parole* (the French word for speech). Twitter as a platform, for example, is the langue and your Twitter account is a parole. Saussure furthermore argued that words do not simply refer to objects in the world for which they stand. Instead, a word is a linguistic sign consisting, like the two sides of a coin, of two inseparable parts: signifier and signified.

A signifier is a "sound-image" (a mental imprint of a linguistic sound); the signified is the concept to which the signifier refers. The landing gear of an aircraft as a phrase in a tech document is a signifier, and as an aircraft instrument it becomes a signified. If the signifier is a wrench, the signified will be the wrench in our imagination that we can picture.

3.1 WORDS IN CONTRACT

It goes without saying that one of the effective ways to communicate with other people is naming objects. We understand each other when talking about a laptop, a special connector in the flight deck or an aircraft fuel pump. Functioning as a meaning system of its own and the tool for using meaning, language serves as a great example of how publicly shared understandings are crucial to the effective functioning of both the culture and the individual. Similarly, an individual who gets pulled back by the language barrier and fails to communicate with others would be destined to a lonely, and most likely miserable, existence.

In his "A Table is a Table" (Ein Tisch ist ein Tisch) written in 1995, Peter Bichsel, the Swiss writer, tells us the story of an old man who wishes to change his life by giving new names to familiar, ordinary things, activities, and so on. Failing to be entertaining enough, playing this game of naming things differently makes him gradually lose his connection to the outside world. For example, the old man called bed "picture," table "carpet," or newspaper "bed." Obviously, after stopping to make sense to the rest of the world who followed a different system for communication, the old man failed to follow any conversation taking place around him. He lost touch with reality as a result of the nonstandard linguistic system he had invented with the only user of that system being himself.

In Bichsel's story, the old man who is the inventor of a new, personal language has to laugh when he hears someone say, "this airplane has been sitting in the hangar for 2 months now" or "these apprentices were supposed to rotate in September." As the Bichsel story illustrates, central to communicating with others is the shared meaning systems. Walking against the grain would come with the severe punishments of stopping to make sense and being totally isolated from the rest of the world. The poor old man began to fear speaking to other people, since it was not only he who could no longer understand them, but they also could not understand what he was saying. He ended up ceasing to speak.

Being innovative in communication would only pay off if you knew how to play the game. Getting too creative with this game could make your identity disappear. Never let your writing take you where your brain didn't arrive at earlier.

3.2 REFERENTIAL VS. NON-REFERENTIAL FUNCTIONS

Among the various semiotic approaches to verbal communication, the one by Roman Jakobson is perhaps the most insightful one. Jakobson's analysis suggests that verbal discourse goes well beyond the function of simple information transfer. It involves determining who says what to whom; where and when it is said; and how and why it is said; that is, it is motivated and shaped by the setting, the message contents, the participants, and the goals of each interlocutor.

It should be also remembered that our awareness of nonverbal communication is vital not only for our survival, but also for understanding the needs, feelings, emotions, and thoughts of others.

Scientific and technical writing constantly refer the reader/user to things outside of their composition. In this case, a centrifugal force that tends to push things away from a central point, in different directions, is created. The elements in a fiction – a short story by Anton Checkhov, for instance – consist of signifiers and signifieds that form on the pages you see and also in your imagination with no concrete, objective reality. Documented procedures in a technical manual are comprised of signifiers, such as words and figures. Their signifiers, or what those words and figures actually refer to in reality, are objective enough for the users of such publications to locate, identify, touch, troubleshoot, modify, remove, or install (for example, think of those instructions or standard operating procedures for working on the ejection seat of an aircraft, or troubleshooting a glitch in your smartphone).

3.3 HORIZON OF EXPECTATION – PARATEXT

The author's name on the cover of a book is a paratextual element that can establish a horizon of expectations which the text itself can confirm or subvert. Revision number, title, writer, department, distribution list, prepared by, reviewed by, approved by, … they all signify functions that determine the reader's (user's) expectations. How many times have you purchased a book only because of seeing your favorite writer's name on the cover, or simply because your favorite publisher published it?

4 To Be Writing or to Refrain From Writing?
That Is the Question

Research shows that writing has existed for a mere 5,000 years – a miniscule amount of time, compared to spoken language. Cuneiform, the oldest known written language, was invented by the Sumerians around the end of the fourth millennium BCE, in the part of the world that is now southern Iraq. Cuneiform was developed as a means for keeping agricultural records. The earliest symbols were actually clay tokens that represented concepts (like bushels of wheat, for example), rather than speech.

Spoken language owns more channels than writing. In other words, speech is provided with more channels for carrying meaning. In speech, there is volume (loud to soft), pitch (high to low), speed (fast to slow), emphasis (present to absent), intensity (relaxed to tense), and, depending on the situation, body language. Used for an indefinitely large number of purposes, writing is used to express feelings, report problems, tell stories, complete forms and checklists, give instructions, retain records, and much more. However, writing is different from those professions that pay you after putting in a day's work – punch in/punch out, do your 40 hours, receive a paycheck at the end of the month, repeat. Writing is a passion, a medium where there is time to think and evaluate, to rethink and reevaluate, to write and rewrite in order to use the components (words) of the system(language) as a way of shaping thought as effectively as possible.

The shared cultural meaning, although invisible, helps us perceive, understand, and interpret the world around us. As a writer we should bear in mind that we are surrounded by the assemblage of meaning systems that affect our thoughts and actions even if we are unaware of its presence, like fish in the water. Writing is much more premeditated than speaking.

Henry Fielding (1707–1754), the English writer, once said that writing is not about setting pen to paper, as many would imagine. Although many may falsely believe that English is an important subject for students of humanities and arts, scientists and engineers spend much of their working lives writing with a pen, a pencil, or using a computer to accomplish the task of writing. I remember that in one of our meetings with a room full of design engineers, along with the regulatory agencies representatives, I was asked by a Designated Airworthiness Representative (DAR) if I had been an English major. Why? Simply because of the corrections I had made to the eight-page long document this gentleman had created to satisfy few regulatory requirements. He didn't sound happy, nor did his tone, but my corrections were accepted by the team and got fully implemented at the end.

Writing is a complicated business. The act of writing can be a choice, but the act of understanding or interpreting would be a challenge, a non-choice, a non-intention. A well-developed sense of purpose and audience is the key to good writing. With no clearly stated purpose or target audience, the worst form of writing gets produced, which violently deviates from the intended purpose behind its creation. What would happen if you systematically deviated from your purpose? Well, many individuals, teams, and groups would end up suffering as a result of poor customer service.

Good writing is nothing but clear thinking made visible, accessible, and memorable.

Mark Twain once said, "The difference between the almost right word and the right word is really a large matter. It's the difference between the lightening bug and the lightening" (Applewhite and Evans III and Frothingham 2003). Do not fill your writing with empty, powerless words.

Writing is never a straightforward linear process. My experience indicates that the best writers never think their first pass is the final one and that's why they continually spiral between drafting, redrafting, editing, proofreading, and research. You choose. Either your writing is finished completely or your reputation will be completely finished.

Writing is about clear communication. It's making sure that people have the information they need to do what they're supposed to be doing. Using standard format is like a road map that elaborates a series of activities being carried out by people in a logical sequence. One of the most obvious, yet rarely discussed, components of writing is deciding whether or not a document should be written at all.

Your writing is ink and paper only, until someone picks it up and reads it. Then it becomes live and directs the reader toward creating your intended world. Without an audience, your singularity would decay.

4.1   LET THERE BE THE USER, AND THERE WAS NO LOSER

We all have been in situations where users were in a state of frustration as a result of having failed to carry out the task by themselves. Since users are the true owners of the documented requirements, when they read your instructions, they need to be able to follow each step easily and clearly. Technical writing, on the other hand, is done with a singular style and format that describes why you are doing the work (introduction); what you did (procedure); what happened (results); what it means (discussion); what was learned (conclusions); and what is to be done with the results (recommendations).

To achieve predetermined, thoroughly studied, well-planned outcomes and objectives, you must first define them, then measure them, then control them, and finally assure them – definition, measurement, control, and assurance being objective phenomena.

Now check out this transmission between an air traffic controller and a pilot:

"Descend two four zero zero feet."

As you see, the similarity (homophony) between "two" and "to" can lead the pilot to understand and interpret 400 feet instead of 2,400 feet. The aircraft would crash into high ground.

In his "A Fable" published in 1909 (The Complete Short Stories of Mark Twain, 2005), Mark Twain tells the story of a human artist who paints a picture, but among the animals, only his house cat can find its beauty. After painting the picture, the artist places it in a way that it's viewed through a mirror, simply to double the distance and soften its beauty. Although the cat explains to other animals that a "picture" is something flat and a "mirror" is a hole in the wall, they end up standing between the picture and the mirror and what they'll perceive is just a reflection of themselves. Ending up being called morally and mentally blind, the cat draws a conclusion, a moral, "You can find in a text whatever you bring, if you will stand between it and the mirror of your imagination."

What is the role of imagination in a, say, technical document?

A manual, a brochure, and a piece of technical publication, all bring together numerous procedures for a common purpose. There is always this triangle consisting of you (the writer/sender/creator), the procedure or instructions (creation), and the reader or user (receiver/co-creator). Never assume your reader knows exactly what you're talking about. Write from a reader's standpoint. Write for them and make your voice heard and acknowledged.

4.2 HOW MUCH INFO WOULD INFORM?

In his short story "Del rigor en la ciencia" ("On Exactitude in Science"), which consists of just a single paragraph, Jorge Luis Borges describes a special country. In this country, the science of cartography is so sophisticated that only the most detailed of maps will do – that is, a map with a scale of 1:1, as large as the country itself. Its citizens soon realize that such a map does not provide any insight, since it merely duplicates what they already know. Borges' map is an extreme case of the information bias, the delusion that more information does guarantee better decisions.

So, too much information is not necessarily good information. Enough information minimizes confusion and eliminates misinterpretation. Writing in accordance with this principle is highly encouraged and strongly recommended.

Daniel J. Boorstin once said, "The greatest obstacle to discovery is not ignorance – it is the illusion of knowledge."

4.3 YOU WRITE, BUT THEY'LL OWN, RIGHT?

4.3.1 IF RIGHT, THEN DO WRITE

Language is a powerful tool lent to us equally in order to get the job of communication done. You do not tell your reader (user) what to do, language does. Situations and contexts play a significant role in facilitating understanding and compliance. You help your readers/users by looking at things from their perspective. That is why the word "ambulance" is printed backward so that it appears correctly (forwards) for the audience – in the rearview mirror of a car – to ensure that the user (driver) can read it and respond as quickly as possible. Technical publications are meant to show the user (reader) how to do things right. They draw the line between what "to do" and "how to do" them, against what "not to do." Writers export; readers import. That's how you do business. No imports and exports are exempt from the explicitly applicable rules, requirements, and regulations. Failing to tell our readers about the water and its characteristics clearly will result in their drowning.

Other forms of writing – say, fiction – on the other hand, ask the reader to question the authenticity of the right thing by creating a parallel universe to the real, familiar one. In this case reality won't be that real and our self-understanding gets constantly and seriously challenged through the work of literature in which connotations and metaphors override denotations. A technical document stands firm against open-endedness. It cannot survive if it opens itself to different interpretations. The life span of your trust by the intended user, the reputation of the organizations who use your writing, and your reputation as the creator would depend on how you say what you wish to say.

The best writing represents you in a way that feels as authentic as a conversation and connects you to an audience beyond your immediate circle. There are different types of writing: expressive, expository, or descriptive writing. All writing is aimed at achieving some purpose or the other. Technical writing, however, is very specifically aimed at achieving certain purposes. A training manual that does not exactly achieve what it is intended to achieve would end up sitting there gathering dust. A well-planned and well-designed documented piece of user instructions has to take into consideration some important factors even before the process of writing begins. Writing is a strange act. It is conceived in the mind, born on the paper, and remembered dearly if welcomed, but perishes instantly if unwelcomed.

Tech writers do not own their creation, they co-own it. Who is the real owner then? The user. The receivers of the message. The more the real owner gets involved, the more effective the documents you produce will be.

4.4   USER GUIDE GUIDES USER

4.4.1   Shifting Focus from "I" to "We"

Communication isn't as simple as saying what you mean. How you translate what you mean (your thoughts) into words is crucial and differs from one person to the next, because using language is a learned social behavior depending on a variety of factors. How we talk and listen are deeply influenced by cultural experience reflected in our narratives. Although we might think that our ways of saying what we mean are natural, we can run into trouble if we interpret and evaluate others as if they necessarily felt the same way we'd feel if we spoke the way they did. Clarity is key, but it won't be achieved by unclear thinking.

It'd be great to encourage individuals to refer to the tech pubs and use them to perform maintenance work on, say, aircraft. However, if you make this a difficult process or a challenge for them of any form, then you end up mass producing documents that are merely physical objects occupying space and time and frowned upon. People give up on the tech pubs if they are not accurate or usable. Make your readers feel important by writing for them – the true owners of your writing. Stick with simplicity in both form and content. When I teach courses such as "smart systems," my students and I hook up and fire up a piece of equipment, such as a smart switch, that has the capability of connecting to the Wi-Fi and being controlled via an application that you download to your smart phone. Then, after having a couple of students play around with the device – disassembling and reassembling – I ask them to go back to their desks and create a user guide. They are tasked to write down clear, concise, and valid instructions to walk the user through the smart device's installation and the firing-up process in less than seven steps. The best one (if any) that is the closest to the way the target audience, regardless of their background, age, nationality, etc., would understand and act upon and always receives additional marks. Being a strong advocate of writing, and to make up for neglecting the significance of writing by their previous trainers and the education system, I also ask my students to write down what they learned on every single day. They end up loving and appreciating what they're doing.

Writing is highly in demand, yet poorly practiced these days. I don't hesitate to grant my audience the opportunity to practice writing and find their voice because I hate to see phones and gadgets getting "smarter."

5 Technical Communication

All writing takes place in a specific context, and all writing involves these elements: a writer/speaker, a message, and a reader/listener. Two of the main differences between technical communication and other types of communication are that (1) its subject matter usually requires some type of specialized knowledge and (2) it provides a bridge for the nonspecialist reader to complete an action successfully.

Technical communication transmits detailed and complex information; both variables must be involved for a document to be considered technical writing. A telephone book, for example, has a lot of details but little complexity. On the other hand, an equation for atomic fission may take up less than a page, but its complexity would make it highly technical. Your purpose in technical writing is to transmit highly complex and detailed information in a format that readers can understand at their end with no difficulties.

Tech writers are cosigners. Technical communication refers to the activity of preparing and publishing specialized information in a way that allows nonspecialists to understand and use the information to accomplish some task and to understand some concepts. A good tech writer leans more toward the "show" rather than the "tell" part of the story s/he tells. Tech writers are humans talking to other humans whose life may depend on the tech writers' selection of words.

Are you aware of how words, tone of voice, and nonverbal messages interact to influence the emotional impact of your communication?

Eight characteristics define excellent technical documents:

- Honesty
- Clarity
- Accuracy
- Comprehensiveness
- Accessibility
- Usability
- Professional appearance (neatness)
- Correctness (adhering to the conventions of grammar)

Technical writing typically requires give and take, a dialogue, a follow-up, an input, and an action. It causes the person(s) at the other end to respond, react, or do something. Tech writing is not a workload (quantity) to be managed; it's a task (quality) to live with. There are three factors very important for technical writing – purpose, audience, and tone. The purpose of writing and the audience very often set the "tone" of a piece of writing.

What is your purpose? A clearly defined purpose would determine the quality of the whole process. Practice this:

Our purpose is to _____.
The user(s) that _____,
so that they will _____
after reviewing this document.

5.1   WHAT IS TECHNICAL WRITING?

"We have created so many tech pubs. Have received no serious compliant, which means we have done a pretty good job" is not a good judgment at all if you have not kept a two-way street – an open channel – to receive feedback. The culture of the organization may encourage no constructive feedback or make it way too complicated for those who would like to make their voices heard.

Good technical writing follows guidelines to ensure that users get the help they need from a document. War mobilizes technology. The aftermath of World War II necessitated the quick dissemination of technical information through the medium of reports. The practicing scientists, engineers, economists, psychologists, and other professional men and women have to keep themselves well posted with this sort of increasing knowledge. Thus, with a view to assorting a miscellaneous type of specialized or technical knowledge or skill, a unique method was evolved by the professionals and it was developed in due course of time as technical writing. It requires specialized knowledge of the subject on the one hand and a cogent and logical exposition of thought in a chiseled strain of language on the other.

Here are some definitions for technical writing by some professors and the subject matter experts.

Technical Writing

- is a specialized form of exposition, a form of discourse distinguished from other forms in certain essential features;
- is expository writing: it belongs to that great class of writing intended primarily to convey information to the reader;
- involves material of a specialized nature, meant for a specific reading group;
- is that branch of writing in which knowledge of a subject in particular is required; and together with knowledge, a succinct presentation, a technical style, and a scientific process are also of great significance;
- is a kind of writing in which the imaginative fervor of a poet, the fictional zest of a novelist, the make-believe credibility of a historian, the magical sway of a storyteller are not required at all.

To technically write right, a technical writer is supposed to be a man of reason – dry reason. Imagination would help only if it was used to help the reader with a better understanding. Readers of the tech writing are also doers who are seriously prohibited to go wrong.

5.2 SELL YOURSELF FIRST, YOUR PRODUCT NEXT

In sensitive contexts, safety and quality go hand in hand. Humans are the most significant, yet most vulnerable, components of any system. The worst form of the human failure involves communication breakdown, a lack of teamwork, and decision-making errors. The human element is the most flexible, adaptable, and invaluable, yet the most vulnerable to influences which can adversely affect its performance. Machines are straightforward because they can work in predictable ways. Humans are infinitely more complex in the way they think and act, therefore it is much harder to analyze their behavior when undesired circumstances arise.

If the tech pub you have created is full of errors – syntactically, semantically, and in design – users will wonder if you were also careless in gathering, analyzing, and presenting that technical information. You'd sound too detached. If your professionalism is doubted by the user, they will be less likely to accept your conclusions or follow the instructions/recommendations prescribed by the mindset (you) behind them.

The rule of thumb of technical writing is to write for the lowest common denominator – at an eighth-grade level. Some useful general guidelines are:

- Use an impersonal language and declarative sentences.
- Avoid using slangs, idioms, and nonstandard expressions.
- Make your writing manageable and structure it through the use of paragraphs and bullets.
- Improve your non-technical or soft skills to communicate more easily, effectively, professionally, honestly, and openly with the team of writers as well as players from your target audience. The ivory tower syndrome could easily and silently form.

5.3 "TECH PUBS SUCK," THEY SAY

5.3.1 NOISE OVERRIDING SIGNAL

The structures made in India and Egypt BCE show evidence of measurement and inspection in the process of cutting and sculpting the stones to construct forts, pyramids, idols, etc. After the United States entered World War II, quality became a critical component of the war effort. Bullets manufactured in one part of the United States had to perfectly match the rifles made in another part. So communication is not a new phenomenon; it is as old as the human civilization. But what happened with their requirements back then? Were they being documented, or verbally uttered? Well, it'd have sucked if they had not been documented.

"The Tech Pubs suck!" is a statement I've heard from many people in the industry. Is that a statement? An excuse? A justification? A complaint? Whatever it is, how could a piece of writing suck by itself? Was it created out of the blue? Did it emerge randomly? Is it set in stone?

After looking at Michelangelo's David, someone asked him, "How did you know he was in there?" Michelangelo replied, "I just took away everything that was not David."

5.4   BRIEF LEAVES NO GRIEF

There is a story about a Broadway producer meeting with an eager young director and playwright. The director-playwright spends forty intense minutes over lunch, breathlessly outlining for the producer a great idea for a play – a play that is sure to be a blockbuster. Finally, in exasperation, the producer turns to the young playwright and says, "If you can't write your idea on the back of a business card, then you don't have an idea."

If you agree with that Broadway producer, then a good writer can be a reductionist, a simplifier, an expert facilitator always looking for that one gem in a bucket of muck. There's that one irreducible nugget that captures an essential truth about life. Organizations depend on groups and teams to do much of their work, but teams often founder on structural flaws. A team's structure may come from the top initially, but often needs to evolve locally to meet the challenges of the game, the work at hand.

6 Mission
Omission of Miscommunication

People communicate with people; machines communicate with people; so do procedures and tech pubs. An Airbus A320 mechanic follows the steps in the Aircraft Maintenance Manual to troubleshoot a system, replace a component, and perform all other necessary maintenance tasks on the aircraft. A Boeing F/A-18 mechanic takes the same steps by exactly following the procedures prescribed in the aircraft tech manuals. A pilot flying an aircraft knows how his engines are behaving; an aircraft painter also has procedures to follow as a process owner in his job.

Since the dawn of time, communication has been an integral part of human life, and the need for better technology to support our communication has been growing continuously. Communication plays a central role in our lives. Your personal success and happiness depend largely on your effectiveness as an interpersonal communicator.

Over the centuries, we have invented many different methods of communication to bridge the barrier of distance. Communication started its journey when our ancestors used pigeons for long distance communication. Later on in 1830, land line communication started its journey. Have you noticed when children speak? They are not talking to anyone in particular, they are just thinking aloud. Beginning to speak, a toddler learns to name things before s/he learns about language itself. So, would this still be called communication? Well, it does lack the two-wayness of communication.

Communication is an art, not a science that encompasses inconsistencies, interpretations, and emotional variables. My experience indicates that communication – composed, delivered, or received – is greatly influenced by every individual's unique frame of reference, life experience, bias, education, and belief system. Although we follow the lines and words in a text with our eyes and listen with our ears, signals still have to pass through filters – already and strongly in place – for "understanding" to take place. Becoming a highly skilled communicator is not be easy. The good news is that the skills of effective communication can be acquired. Though some of us are inherently better communicators than others, we all can learn and improve ourselves.

As a technical writer, other people's happiness, success, challenges, miseries, gains, losses, and struggles would directly depend on your craft and how you communicate.

6.1   ENGLISH AND AVIATION

In 1951 the International Civil Aviation Organization (ICAO) instated English as the official language of aviation. The ICAO advised all airports and routes to operate in their native language, but to use English for international flights. ICAO Document 9835 that came out in 2004 is a Manual on the Implementation of ICAO Language Proficiency Requirements (LPR) for the airmen. Although pilots and air traffic controllers must use standard phraseology to communicate, they must also be able to handle the out of sequence communication in situations that are not prescribed in the radiotelephony manuals – bees attacking pilots on the flight deck, for example. This form of communication – being 100% verbal in nature – has proved itself crucially significant.

The ICAO has offered a rating scale as part of the Language Proficiency Requirements (LPR) delineating six levels of language proficiency ranging from pre-elementary (Level 1) to expert (Level 6) across six skill areas of linguistic performance: pronunciation, structure, vocabulary, fluency, comprehension, and interactions.

6.2   WANTED: A CULPRIT NAMED LANGUAGE

> KLM to Tower: "Ah roger, sir, we're cleared to the Papa Beacon flight level nine zero, right turn out zero four zero until intercepting the three two five and we're now at take-off."

> **– From the Cockpit Voice Recorder on the day of the Tenerife Disaster (1977)**

Verbal communication could have weaknesses. "Send reinforcements, we're going to advance," could end up being communicated/interpreted as "Send three and four-pence we're going to a dance."

The human element is the most flexible, adaptable, and valuable part of the system, but it is also the most vulnerable to influences which can adversely affect its performance. James Reason's great metaphor of the Swiss cheese model captures how errors emerge. Considering each of a system's layers – from the equipment, to the human, to the work environment, to the coworkers, to the company's culture – disaster occurs only when the holes in the Swiss cheese line up. Wouldn't language and human communication – hazards to be seriously taken into account – be considered as layers of defense in this model?

We must always keep in mind that there is a chain of events leading to accidents and catastrophes. We won't be able to pinpoint one specific cause (Figure 6.1).

Over 800 people lost their lives in three major accidents. A common contributing element: miscommunication. Air accidents have been tied to language competence. Let's look at three of them and briefly talk about each here.

The world's worst Pan Am airline tragedy occurred in 1977, when a KLM Boeing 747 collided with a Pan American Boeing 747 on a foggy runway at Tenerife, in the Canary Islands, killing 583 people. Seventy-seven people survived aboard the Pan Am aircraft. Determined to proceed with an unauthorized takeoff, Captain van

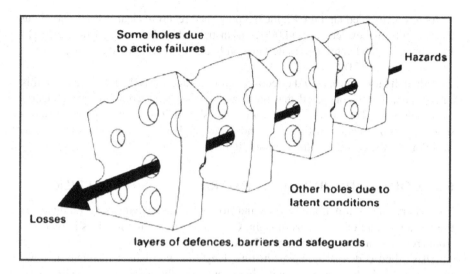

Some holes due to active failures

Hazards

Other holes due to latent conditions

Losses

layers of defences, barriers and safeguards

FIGURE 6.1   J. Reason's Swiss cheese (accident causation) model.

Zanten (the KLM pilot) turned the throttles to full power and roared down the same foggy runway where the other aircraft – Pan Am Boeing 747 – was taxiing. The Dutch captain's vague "We are now at takeoff" was interpreted by the controller as "We are now at the takeoff position." What the Dutch captain meant to say, however, was "We are now actually taking off." The statement the captain made in English was an unusual phrase in standard aviation phraseology.

Avianca Flight 052, a Boeing 707 from Bogota (Colombia) to New York (JFK), experienced a loss of power to all four engines and crashed approximately 16 miles from the airport. Seventy-three lives were lost. In this preventable crash, Flight 052 was placed in holding three times by the air traffic controller for a total of 1 hour and 17 minutes. Finally, suffering from fuel starvation, the pilot contacted the control tower again, *"Avianca zero five two we just ah lost two engines and we need priority please."* The air traffic controller, acknowledging their message and the 15 words in it, radioed back, *"Avianca zero five two turn left heading two five zero intercept the localizer."*

Running out of fuel, the aircraft crashed. The flightcrew, failing to use the standard phraseology to properly declare an emergency by picturing their situation clearly for the air traffic controller, used the word "priority" which means "attend to me now" or "first" in Spanish. When you think in your mother tongue and speak in another language, then your thoughts get lost in translation.

The Charkhi Dadri mid-air collision occurred on November 12, 1996, when Saudi Arabian Airlines Flight 763, a Boeing 747 en route from New Delhi (India) to Dhahran (Saudi Arabia), collided in mid-air with Kazakhstan Airlines Flight 1907, an Ilyushin Il-76 en route from Shymkent (Kazakhstan) to (New Delhi), over the village of Charkhi Dadri (Haryana, India). With no survivors, all 349 people were killed in this deadliest mid-air collision in history.

As it was revealed later on, Flight 763 pilots had received clear instructions about their altitude and leveled off at 14,000 ft, as instructed by the traffic controller. Flight 1907, however, had continued to descend, although instructed by the traffic controller to level off at 15,000 ft.

Although lack of proper and effective communication was the major cause of this catastrophe, it was concluded that the flightcrew were not communicating clearly with each other. The radio operator on the Ilyushin Il-76 – the only person responsible for communication with the ground – was not being paid attention to by both the pilot and the co-pilot of the doomed flight.

6.3   CODE-SWITCHING: THE POWERFUL, INVISIBLE BARRIER

Vocabulary and grammar are learned and stored in long-term memory. Language is used as a medium of conscious thought. Communication often involves individuals from cross-cultural and multi-lingual backgrounds.

Also called codemixing, code-shifting, language alternation, language interaction, or linguistic shifting, code-switching refers to conditions in which two linguistic systems, while remaining separate, are brought into contact. Thus, the code-switching phenomenon presumes the existence of the language independence phenomenon.

Change of language from one sentence to the following and starting a new topic or an utterance in a given language then switching to the other and continuing utterance in the other language, code-switching has been hard at work. What is very interesting about code-switching is that most speakers don't even notice when they are changing tongues. It is likely that this is due to the speaker being more focused on expressing an idea, and in the process of formulating an accurate expression of that idea, they make use of the known vocabulary in the two languages.

For example, English is generally considered to be a language with a relatively fixed word order. In practice, this means that the subject **S**, the main verb **V**, and the object **O** have fixed positions in relation to each other. English follows the order **S-V-O** order while, say, Farsi follows the **S-O-V** order.

Another example is about nouns. All nouns are countable in Farsi, and the subdivision of "countable vs. uncountable" does not exist in this language – e. g. I am learning how to operate these *equipments*; she spilled all the *coffees* over the *moneys*; he does have *a* good *news* for us.

Another difference is the lack of "the" in Farsi, as a definite article in English. In other words, a definite generic noun lacks a surface structure sign in Farsi – e.g. dog doesn't like fish; teacher asked us a question; pen is on table.

Wouldn't a phrase like "my furniture" signify that language is capable of counting the uncountable?

I am 100% sure that all immigrants have experienced the unique phenomenon called code-switching in their new country. One personal experience to share is from one of my jobs as a supervisor managing 60+ employees from all walks of life. Although my learning curve was quite steep, I often could not separate myself from the powerful, hidden power of culture and the first language.

Me (to one of my employees, a native American): "I need you to go to location
    A right now."
Him: (Puzzled. Pause. Silence.) "Well, you cannot talk to me like that."

Then, after inviting him to a drink that afternoon, he told me that I had
sounded authoritative and he had been irritated by my selection of words to convey a significant, time-sensitive message. My superiors would tag this employee's behavior as "insubordination" and I could have written him up, but I learned
another lesson. Although I had gone to the best English school back home, rose
to the top of my class, and graduated with honors, I had not been taught these
things properly. Why not? Well, to eliminate these culture shocks and hiccups for
yourself and others, you must live in the country of your desired target language
for a while.

In the supervisor–employee anecdote above, on the contrary, this would have
been quite a standard, normal statement in a similar situation in my first language,
with no annoying consequences. The system of a language is at once cognitively
based and socially constituted, being a mental resource for individuals, yet rooted in
the social life of communities of speakers.

6.4 BAD WRITING: LATENT AND ACTIVE FAILURES

A primary goal of communication is to be understood. Coming to an agreement with
the addressee – reader or listener – is rooted in making a choice. While we cannot
control another person's agreement, we are in control of presenting clear ideas in an
effort to be understood.

James Reason's "Swiss cheese" model of mishaps, as discussed earlier, tells us
that the human contribution to complex system breakdowns and catastrophes are not
caused by one event in isolation, rather by a number of causes and conditions that
line up. Reason defines these failures as active and latent. Active failures are unsafe
acts committed by the frontline people who have direct, active roles in the system.
Latent conditions are preconditions for unsafe acts.

Bad writing or useless documents, being active failures, do create latent failures
soon. Very soon. Unsafe acts can be committed by you; unsafe conditions can be
created for you. Poorly created instructions, guidelines, and procedures fall under
both.

There's a famous saying that goes: planning without action is futile. Action without planning is fatal. By talking more than necessary – a skill not hard to attain – we
dig up holes we will fall victim to as writers. You'd be better off telling your reader
one good story than bringing 1,001 reasons for them to read your story.

To write in English – a foreign language for me – I'd have to look up the words in
an English to English dictionary and look at examples carefully in order to achieve
the right anglification. At first before mastering the technique that I just mentioned,
I wrote in English and partly in Farsi, the two languages being mingled in the most
wonderful yet strange manner.

6.5 LOST IN TRANSLATION

It is recorded that Giovanni Schiaparelli was working with a small refracting telescope in an observatory in Italy, where he drew remarkably detailed maps of Mars by observing the planet night after night. The telescope he was using was not very big for observing far-off planets. When Schiaparelli combined his sketches of the fuzzy images, the planet went from a hazy disk to a detailed globe, covered with surface features, which were connected by lines that he called "canali," the Italian word for "channels."

Schiaparelli published some of his best work between 1877 and 1886 in which his canali was translated to the English-speaking world as "canals."

"Channels" can occur in nature, but "canals" are built, presumably by intelligent beings. One word can create all the difference. Select your words before misunderstanding selects you.

6.6 MISCOMMUNICATION COMMUNICATES LOSS AND COST

Have you ever talked with zero communication taking place? Have you been to meetings in which the CEO was lecturing and bragging about his accomplishments in marketing and sales when the director of marketing sitting in the same meeting with the CEO looked puzzled and confused? I have. Communication does imply challenges and roadblocks. Poor communication or miscommunication may undoubtedly be tracked down as one of the root causes of any failure.

NASA had a disastrous experience with the Mars Climate Orbiter, which burned and broke into pieces in September of 1999, after months of traveling to Mars. This $100+ million robotic space probe malfunctioned and was destroyed because of catastrophic miscommunication. NASA navigators were under the assumption that a subcontractor was using metric measurements when in fact English units had been used. NASA had converted to the metric system 3 years earlier (1996), and this change of standard was not apparently and effectively communicated to all involved parties, stakeholders, and organizations who were directly or indirectly involved in this mission. This monumental miscommunication resulted in a huge financial loss, reputation, waste of countless hours of valuable effort by engineers and decision makers, and the loss of critical data the mission was intended to generate.

Everything ultimately becomes obsolete. Change, being scary, is essential, but communicating it and asking for feedback do transcend any hypothesis.

6.7 POWERPOINT: POWERFUL BUT POINTLESS –
THE COLUMBIA DISASTER

Did PowerPoint make the space shuttle crash? This tragedy occurred some 18 years after the Challenger accident (1986), this time not by O-rings. The Columbia lifted off on January 16, 2003, for a 17-day science mission. Upon re-entering the atmosphere on February 1, 2003, this orbiter suffered a catastrophic failure due to a breach that occurred during launch when a 2.5 pound falling foam rubber insulation struck the underside of its left wing, as per the NASA investigators. The orbiter disintegrated and all its seven crew members were killed approximately 15 minutes before its touch down at Kennedy Space Center. But why?

Dr. Edward Tufte of Yale University, a researcher and an expert in information presentation and communications failures, studied how the slides used by the Debris Assessment Team in their briefing misrepresented key information. Besides criticizing the sloppy language on a particular slide, Tufte claimed that an overloaded and misleading PowerPoint slide which was supposed to indicate gaps in the shuttle's safety system was one of the causes of this catastrophe in the chain of events. The very slide, according to Tufte, contained a disorderly list of bullet points, and it was the intensely hierarchical bullet list that led people to ignore the hazard and mitigate the risks.

Acceptable design, besides all other benefits, contributes to an acceptable level of safety.

Information is raw material cooked by language, served with language, taken by language, and digested by language.

6.8   USE A CHECKLIST TO MAKE YOUR CHECKLIST

Poorly designed checklist murdered 156 people when Northwest Airlines Flight 255, a McDonnell Douglas MD-82, crashed on August 16, 1987, after liftoff from Detroit Metropolitan Wayne County International Airport. Why did it crash? The flight crew's failure to use the checklist to ensure the flaps and slats were extended for take-off was the main cause in the chain of events which in turn was attributed to failure to follow every step of the pre-flight checklist. With 156 fatalities, including 2 on the ground, aircraft manufacturers and the airline management learned the importance of design in checklists and flight-deck procedures to foster item completion. The flight crew did not discover that the airplane was configured improperly for takeoff. With flaps and slats not properly extended before takeoff, the first line of defense had already been breached. The aircraft had been built perfectly, but it crashed.

As a result, McDonnell Douglas (manufacturer of MD-82) and U.S. Federal Aviation Administration (FAA, the regulatory agency) responded with appropriate actions. On September 1, 1987, McDonnell Douglas issued a telex to all DC-9-80 operators recommending the airplane checklist be changed to include a check of the takeoff warning system prior to departing the gate on each flight.

In April 1991, "The Use and Design of Flightcrew Checklists and Manuals" was published with a long list of recommendations while placing the blame on the checklist design.

6.9   LEFT, LEFT OUT; THAT'S NOT RIGHT – FLIGHT BD092

Poor interface design for users can be the root of all evil. Who is design meant for? In the British Midland Airways flight BD092 accident in Kegworth on January 8, 1989 – the worst air accident in the UK in many decades – badly grouped and poorly displayed engine instrumentation in a Boeing 737-400 cockpit led to a catastrophic visual selection error, contributing to the flight crew shutting down the wrong engine. When the captain asked his first officer about the faulty engine, the latter, after some hesitation, replied that it was the No.2 engine. Investigators determined that the pilot apparently shut down No.2 engine (right engine), even though the plane was having trouble with its No.1 engine (left engine). But why? Well, the short answer is that the poorly designed cockpit layout – the user interface – resulted in 47 fatalities and many injuries.

6.10   TIE IT RIGHT TO LANGUAGE BEFORE LANGUAGE TIES YOU UP

A DC-8 was carrying cargo for the U.S. Air Force on September 8, 1973. The captain is the boss and the subordinate is his co-pilot. They are flying an approach into Cold Bay, Alaska. The airport is surrounded by mountains, and there are only two safe routes in. Both pilots know this, and both are looking at the same approach chart. But the captain, who is flying the plane, is off track. Looking out the window won't help, because it's dark and they're in the clouds.

World Airways Flight 802 was destroyed when it crashed into Mt. Dutton, near King Cove, Alaska. The co-pilot, failing to be assertive by using language in the right way, preferred to use a rhetorical question to communicate indirectly, "We should be a little higher, shouldn't we?" as captured on the cockpit voice recorder. It is obvious that the co-pilot had been worried, but he did not communicate that. Although this is a cultural thing deeply ingrained in the organizational culture, as we will deal with it later on in this book, a more direct, yet respectable, tone could have prevented this accident.

7 Fair to Err?

It goes without saying that no industry or aspect of life is found to be immune to error and its consequences. Besides, achieving a zero risk level would be wishful thinking. It is not going to help you much by just focusing on human errors, or an equipment failure, without taking into account the socio-technical system that helped shape the conditions for people's performance and the design, testing, requirements, and fielding of that equipment. System accidents result not from component failures, but from inadequate control or enforcement of compliance-related constraints on the development, design, and operation of the system.

Errors may be rule-based, skill-based, or knowledge-based. As we have learned, active failures are those noncompliant acts by personnel that are likely to have a direct impact on the quality of the system. Latent conditions arise from strategic and other top-level decisions.

According to cognitive psychologists such as Don Norman, human error is the cause. But it's also the effect of a bigger cause: bad design.

Unsafe act events can relate to lack of procedures, lack of communication, deficient personnel training/competence, and wrong/inadequate risk perception. Effective communication, in my mind, has to be seriously considered as a vital equipment and an effective barrier between failures and catastrophes in the chain of events. The role of writing cannot be underestimated.

7.1 DESIGNS VS. HUMANS

7.1.1 DONE BY DON

In his *The Design of Everyday Things*, Don Norman writes,

> Today, we insist that people perform abnormally, to adapt themselves to the peculiar demands of machines, which includes always giving precise, accurate information. Humans are particularly bad at this, yet when they fail to meet the arbitrary, inhuman requirements of machines, we call it human error. No, it is design error.
>
> **(Norman 2013)**

Human-centered design (HCD) is the process of ensuring that people's needs are mostly, if not fully, addressed and met. That the resulting product is understandable and usable, that it accomplishes the desired tasks, and that the experience of use is positive and enjoyable.

Norman observes that when users don't understand a product or make mistakes using it, they have a tendency to blame themselves. But would that really be their fault when the product has been poorly designed? For every confusing design element, there exists a better design that reduces or prevents confusion for its target users, and an alternative that is self-explanatory. The great, possible task of finding them is the designer's responsibility.

When developing a design, *affordance* is a property of objects. For example, a button affords pushing, a switch affords flipping, a doorknob affords twisting, an Internet hyperlink affords clicking, a board affords writing on, a slot affords inserting, and a tech document affords reading and following. When the affordance and function are aligned, we generally have a good design. During the design process, the connections between affordances and functions need to be fully explored to ensure the development of user-friendly designs that can be operated conveniently and with minimum learning, instructions, or supervision. Have you ever wondered which shaker contains the salt and which one the pepper? Among the creative solutions with shakers is having holes on the cover in the pattern of S (for salt) or P (for pepper). The famous phrase "Righty tighty, lefty loosey," which can be translated as tighten by turning to the right (clockwise) and loosen by turning to the left (counterclockwise), works for screws, nuts, light bulbs.

We do not need to learn anything about a well-designed door, for example, before being able to open it. The knob affords turning and turning the door will result in opening it. However, can't we recall examples of poorly designed things that we need to learn how to use? There are doors that have handles that afford pushing and pulling. There are also sliding doors which are supposed to slide sideways. People often end up putting many signs on the door to help you and me to learn how to open and close them.

Human-centered design process does require planning.

7.2 WHO OVERRIDES WHAT – AN EXPECTATION CALLED AUTOMATION

7.2.1 A Kingdom Noticed Seldom

On July 20, 1969, Eagle (the Apollo 11 Mission) began its descent. The astronauts were aiming for a place on the Moon called the Sea of Tranquility, which was thought to be flat, with no rocky outcroppings. However, Armstrong and Aldrin, the capable individuals aboard the lunar module Eagle, realized they were off course. The spacecraft's computer was overshooting their original target and was going to land in an area with dangerous boulders and uneven ground. Armstrong, a qualified decision maker, turned off the automatic targeting and autoland features, took manual control, and landed Eagle in a safe place. Mission controllers back on Earth at NASA's Mission Control Center in Houston, Texas, held their breath as they waited for a radio signal from Armstrong. Finally, they heard him say, "Houston, Tranquility Base here. The Eagle has landed." If automation is an empire, humans – the conscious intelligence – are still the emperors.

7.3 EITHER DESIGN RIGHT OR RESIGN RIGHT NOW

7.3.1 THE UI AND THE HMI

Before and after getting up in the morning, we are surrounded by hundreds of systems, products, services, and designs that we need to use on a daily basis. These range from alarm clocks, phones, cars, washing machines, ATM machines, etc. Although instructions are often given to the users to enable them to correctly and safely use these systems, still there are many ways to operate them incorrectly, because of the human factor involved. Considering technical systems from an ergonomic point of view, human-machine interfaces (HMI) have to be designed according to the user's capabilities, taking into account perception as well as cognition of the user interface.

It goes without saying that users are outnumbered by user interfaces. A multimeter can be used by many. But what is this thing called *user interface* (UI)? It is simply the part of the system that enables interaction and serves as a bridge between users and the system. The screen and buttons on an ATM machine, the button/touch interface of a vending machine, switches on the overhead panel in the flight deck, your computer monitor, your smartphone screen, or the layout of a dashboard (speedometer, tachometer, engine and fuel gauges, warning lights, etc.) are examples of a user interface.

When you use the controls on the panel of a dishwasher, the controls form the interface between you and the machine. You are not concerned with the underlying mechanical activities, the technology, or the software of the dishwasher itself. Being easy enough to understand and use would suffice for you as a user to get your dishes washed.

What would any design be good for if it were not meant and built for the user? Design is concerned with how things work, how they are controlled, and the nature of the interaction between people and technology. In *The Design of Everyday Things*, Don Norman talks about the concepts of "knowledge in the head" and "knowledge in the world." The former refers to the amount of learning that we need to have (in our heads) prior to being able to operate a given product or system. Norman argues that a good design can result in products and systems that are intuitive and easy to use by making the knowledge needed to operate these products or systems easily available in them and their surroundings – knowledge in the world.

7.4 LANGUAGE: THE PRESENT ABSENCE

Said doesn't mean heard. Heard doesn't mean understood. Understood doesn't mean agreed.
Agreed doesn't mean retained. Retained doesn't mean applied.

Language and absence go way back. In order for communication and meaning to come into existence, language is devised so that concepts and experiences we are willing to share with others (readers, users, listeners) deem possible. A silent communicator – in verbal and nonverbal forms – won't be left with much luck. Between the addresser (speaker/writer) and the addressee (listener/reader) exists the context. When you immerse yourself in the act of writing, your READER is absent, unless you're writing to yourself. When your reader is reading your writing, YOU will be absent. To me, the act of writing does involve this absence–presence binary, all the time. That would be up to you to make this reader–text relationship an enjoyable, peaceful coexistence or turn it into flames between two adversaries.

Following the act of writing – during which the writer is as present as language and words – acts of reading, interpreting, understanding, and implementing take place. The creator of a standard operating procedure (SOP) would not necessarily need to be present during these stages, nor do his intentions.

7.5   THE LANGUAGE INTERFACE

7.5.1   AN INTERVENTION CALLED INTERPRETATION

In *Alice's Adventures in Wonderland*, Humpty Dumpty tells Alice, "When I use a word, it means just what I choose it to mean – neither more nor less" (Carroll 2009). But, my fellow tech writers do not take Humpty Dumpty's advice too seriously and refrain from following his advice, as you cannot confine words to simply mean what you wish them to mean.

Words do not necessarily represent us and our intentions. In language, we use codes to express thoughts.

Language is much more slippery and ambiguous than we realize. If I put my QA hat on, I should say that a healthful, process-driven, results-oriented organization is founded upon good documentation. If requirements are poorly documented, then they will also be poorly implemented. An undocumented requirement would be a rumor. Documenting requirements is a process that has to follow certain requirements.

With language, being a system, words constitute its design. When you are browsing through the tech pubs, what are you exactly looking at? A user interface called language.

Language facilitates understanding, but it can also facilitate misunderstanding. If not properly utilized, words can undermine their whole purpose of being born, and meaning gets lost in translation. The possibility of full transparency can be rejected as long as words and language are used by us to create a design. Interpretation, then, becomes a creative intervention.

Part II

---

*Language, Would You Mind
Stop Speaking for Me?*

8 Simplified English Simplified

Simplified English is designed to:

1. *standardize* and *reduce* the number of words used, i.e. avoid using more than one word with the same meaning and defining one meaning for each word used; and
2. *standardize* and *simplify* the syntax and grammar used, in order to make maintenance texts clearer and simpler.

The purpose of simplified English is to make their already complicated life easier for the writers (technical editors) and users (technicians and mechanics) of maintenance documents. The team should review for correctness, consistency, accuracy, and completeness.

- Correctness: Is the draft copy using standard English in spelling, punctuation, and grammar? Does it follow the guidelines published in dictionaries and grammar guides?

  When I was a graduate student in the United States, I remember a classmate of mine, born and raised in California, once sent me a text message with three words that I had never seen as an English major: "idk," "lmao," and "chkd." Thanks to Dr. Google, I was able to break the code and respond to him. Spelling may also cause problems, something which mother tongue speakers have difficulty with, never mind learners from another country.
- Consistency: Is the draft copy presenting terms, numbers, and words consistently?
- Accuracy: Does the content sound accurate? Do the tables have accurate numbers? Do equations calculate? Are titles properly used? Is the document telling the truth or does it contain signifiers whose signified is absent?

  As an English teacher, I should mention that those students who are much more interested in accuracy than fluency are often very good when writing.
- Completeness: Does the draft copy contain every single part required for a manuscript of this type? Have all the questions pertaining to tasks been answered? Are all topics being addressed?

The order of words in technical English is very important. Technical English uses a lot of compound words or "noun clusters" that is a chain of words. For example:

door lever
fuel tanks
connector repair
part number
ground servicing operations
left forward passenger door
nose landing gear uplock box
aft cargo compartment door
proximity detector
angle of attack
mean aerodynamic chord
outer RH flap track fairing
attachment bolt heads

8.1   DEEP STRUCTURE AND SURFACE STRUCTURE

According to grammarians there are four types of sentences:

- *Simple* (containing a single independent clause, i.e. one subject and one verb)
  Example: (1) He fixed the problem. (2) All apprentices and artisans graduated last week.
- *Compound* (containing two independent clauses)
  Example: The mechanics are working in hangar 25; the apprentices are on their way.
- *Complex* (containing one independent an one or more dependent clauses)
  Example: When refueling is in progress, you should refrain from smoking around the aircraft.
- *Compound–complex* (containing two or more independent and one or more dependent clauses)
  Example: (1) They planned to go to the skills lab, but they couldn't until they had their PPE on. (2) When request for corrections had been received, the team got to work and submitted them to the technical committee for final approval.

Language has been described by psycholinguists as having a surface structure and a deep structure. When you read or hear speech, you are moving from the surface structure to deep structure – from the way a sentence sounds or looks to its deeper level of meaning. The syntax of a language provides the rules for ordering words properly. A sentence's deep structure refers to the underlying meaning of the symbols (and signs) combined. Pay attention to those structures in the following examples:

The police must stop drinking after midnight.
They are hunting dogs.
I can see the man with a telescope.
She's reading on the train.
Sam watches the birds over the bridge.

8.2 SEMANTICS

Semantics is the study of the meaning of words and sentences. When you ask a friend "How did you do on your exam?" and the reply is "I nailed it," you know that your friend is not saying "I hammered my test to the desk with a nail." Those who are familiar with English do not interpret this expression literally. On the contrary, a beginner English learner or a nonnative English speaker might find such expressions perplexing.

8.3 GENERATIVITY

Generativity means that the symbols of language can be combined to generate an infinite number of messages that have novel meaning. Having only 26 letters, the English language is capable of combining into more than half a million words, which, when combined, can create a virtually limitless number of sentences. Thus, you can create and understand a sentence like "What did the tiger sitting on my birthday cake eat for dessert?", although you are unlikely to have heard anything like this before.

8.4 DISPLACEMENT

Displacement refers to the fact that language allows us to communicate about events and objects that are not physically present. In other words, language unbinds us from focusing only on objects and events that are right before us, here and now – past, present, future, people, things, etc. You can even discuss completely imaginary situations, such as fairytales, using language. You can make up a story, or even change the sequence of events. Language is capable of forging reality. The same language that tells the truth can also lie.

9 Writer
The Deliverer

When you're riding a horse or walking your dog, who is supposed to remain in control? Writers use the vehicle (horse), load it with their desired shipment (policies, procedures, etc.), and dispatch it to the next stop to pick up the recipient (reader), and take them to the next stop to help them accomplish the task(s) and get the job done.

The horse is expected to reach its planned stops and destination safely. The quality of the shipment does determine the end user's satisfaction. I would like to use an example here. In a professional context, people are assigned tasks and duties according to their qualifications (hopefully) to become process owners. These precious assets – the liveware – can be compared to the links of a long chain that forms the system in an organization – yes, true, being a weak link could threaten the strength of the whole chain.

On December 19, 2011, during the holiday season, a security video captured a delivery man – the last link in the big chain of the supply chain management – delivering a boxed computer monitor for a customer. He came up to the house with a package and threw the package over a fence onto the porch. The customer happened to be home and made a video of the shocking event that went viral immediately. Did the delivery guy deliver the package? Yes, he did. But how? It's not *what* we do, it's always *how* we do things. Well, this makes me think of the job of an instructor, a pilot, or a technical writer. While too many activities and steps go into place and take place prior to the delivery phase, it's the frontline people who can mess everything up and damage your reputation if they deviate from the protocols.

The delivery company has been smart enough to pay attention to the lessons learned by taking proper corrective action to take the water hose to the fire rather than the smoke. They resolved the issue with the customer, took disciplinary action against the relevant employee, and have been using that video as a training tool for their personnel.

Are good writers born, or made, or both?

In his *The Sense of Style*, cognitive psychologist Steven Pinker writes:

I'd be the last to doubt that good writers are blessed with an innate dose of fluency with syntax and memory for words. But no one is born with skills in English composition per se. Those skills may not have come from stylebooks, but they must have come from somewhere. That somewhere is the writing of other writers. Good writers are avid readers. They have absorbed a vast inventory of words, idioms, constructions, tropes, and rhetorical tricks, and with them a sensitivity to how they mesh and how they clash. This is the elusive "ear" of a skilled writer – the tacit sense of style which every honest stylebook confesses cannot be explicitly taught. Biographers of great authors always try to track down the books their subjects read when they were young, because they know these sources hold the key to their development as writers.

(Pinker 2015)

9.1 GUN WITH THE WILL

Anton Chekhov is believed to have said that if you introduce a gun in the first act (setting up a problem at the beginning), then by the end of the second act it should have gone off. Application of elements and utilization of objects must have justification with parts serving the whole, and the whole serving the parts. This principle has since become known as "Chekhov's Gun." As a writer, you must constantly bear this in mind that if there's no justification for a single word, sentence, or phrase, or even a word, then they'd need to come out during the editing/proofreading process. Words do not say things, they DO things. In order to better communicate, words must be treated and used better. Not only does the proper use of words count, punctuation also does have a role to play.

Let's look at the following sentences:

- The soldiers, who were sleeping, were killed.
- The soldiers who were sleeping were killed.

In the first sentence, we're told that they were all killed. In the second, only the ones who were not sleeping survived. Punctuation does make a difference, too. Just a heads up, fellow writers.

In his *Syntactic Structures*, Noam Chomsky writes,

> The notion "grammatical" cannot be identified with "meaningful" or "significant" in any semantic sense. Sentences (1) and (2) are equally nonsensical, but any speaker of English will recognize that only the former is grammatical:
>
> 1. Colorless green ideas sleep furiously.
> 2. Furiously sleep ideas green colorless.

(Chomsky 2002)

9.2 AN SOP FOR SOPS

9.2.1 Requirements Required for Requirement

Every communication interaction is unique in terms of purpose, context, mode of communication, and audience (people) involved. When communicating technically, I try my best to tie requirements to some need. If there's no need that has been proactively identified for the requirement(s) to address, then the birth of the requirements(s) would be illegitimate. Documentation may guide performance on new or unfamiliar tasks, but as people become more familiar with a task, they are less likely to refer to the paperwork. This has its risks, particularly if procedures change. By showing people how to do things, we are taking advantage of a possible function of language to fulfill the purpose at hand – giving clear instructions and persuade our audience to perform a task or a series of tasks.

Policies show us *what*, while procedures show us *how* to do things. A standard operating procedure (SOP) needs to be carefully reviewed to ensure it is complete, clear, concise, and valid in order to ensure full compliance and avoid further finger-pointing that would bog the organization down.

In the position description of a tech writer we read: "Able to write clearly." Well, this is an unclear requirement and a tough qualification to ace. How could you write clearly if you were thinking unclearly? I am sure that you, like me, have run into procedures that were open to interpretation. If you pursue synergy, then practice it in the energy of your writing.

In my experience, when creating procedures for tech people, the creator (co-owner) should help the user (owner) through:

- identifying the major tasks and separating them into subtasks;
- writing a series of steps that walk the user through each subtask (often presented as a list of bullets);
- not having two steps in the same sentence or bullet; and
- only giving the reader information at the exact moment that they need it ("just in time" information).

We are working in a global environment and will often communicate with colleagues, supervisors, and customers from different cultures who speak English as a foreign or a second language. Having said that, it would be great to follow a set of guidelines when creating documented requirements:

- Use basic vocabulary that is familiar and understood by readers with average English skills.
- Substitute less frequently used words with more common synonyms. Avoid the use of contractions.
- Use the active voice, which is easier to understand than the passive voice.
- Avoid slang, jargons, and idiomatic expressions that the reader may not understand. Some phrases in American culture have a meaning different from the literal meaning (for example, "get a handle on," "I am all ears," "drop the ball"). Nonnative speakers may be confused by their use.
- Be aware of units of measure, money, and time. Don't assume your reader measures distance in miles or feet instead of kilometers or meters.
- Don't assume the reader purchases items in dollars or buys liquid products in gallons instead of liters. You may need to provide conversions for these units.
- Express only one idea in each sentence and use a simple sentence structure. Long, complicated sentences often can convey that you are not clear about what you want to say. Shorter sentences demonstrate clear thinking and easily convey complex material by creating easy-to-process units of information. Try to keep your sentence length shorter than 15 words.
- Remain as close to the target audience's requirements as possible. Consult and involve them.

When creating a new procedure, ask yourself the following questions:

- Who uses the procedure?
- What is the procedure for?
- When is the procedure used?
- Where is the procedure used?
- Why is the procedure needed?
- How is the procedure used?

Effective operating and maintenance procedures provide a win-win opportunity for both facilities and individual employees. Here are just a few of the recognized benefits of effective, written operating and maintenance procedures. They:

- provide a record of approved, safe operating, and maintenance practices;
- provide consistent information to all users;
- remove guesswork;
- support employee experience and knowledge;
- enhance employee performance;
- document and build upon your facility's experience and practices;
- assist in adhering with industry initiatives and regulations;
- lead to more efficient operations; and
- provide the tools for an effective training program.

9.3   BAD IS NOT GOOD

Good technical writing follows guidelines to ensure that users get the help they need from the tech pubs – a document, a checklist, an SOP, etc.

Compare the following:

Example 1: Click the Picture option in the menu tab under Insert.
Example 2: In the menu tab, click Insert, and then click Picture.

In the first example, the actions appear in reverse order to how they are actually performed. A user must read to the end of the sentence before they learn what action to perform first. The second example is preferable because the actions are listed in the order in which they are supposed to be performed.

In his address at Rice University on nation's space effort in Houston (Texas) on September 12, 1962, President Kennedy said:

> We choose to go to the moon in this decade and do the other things, not because they are easy, but because they are hard.

(Logsdon 2010)

Now check out this version with a slight difference:

> We don't choose to go to the moon in this decade and do the other things because they are easy, but because they are hard. ("Not" sticking to the auxiliary verb "do" would weaken its strength.)

WRITER'S CHECKLIST – WRITING INSTRUCTIONS

- Use the imperative mood and the active voice.
- Use short, easy-to-understand sentences in simple present tense.
- Do not use jargons and technical terminology that your readers might not be familiar with.
- Spell out abbreviations and acronyms. Provide definitions for them in a glossary, footnote, etc.
- Try not to use two or more different words for the same concept. Be consistent.
- Eliminate any ambiguity or vagueness by asking for multiple readings by different individuals. Use appropriate, standard, clear, and concise safety warnings and cautions.
- Tell them what. Show them how. Answer why.
- Ensure and verify that measurements, distances, times, and relationships are accurate and precise.
- Conduct test by asking a member of your audience – the end user – to go over your documented instructions and follow them. Observe them while they perform that.

9.4 TELL THEM WHAT, SHOW THEM WHY

9.4.1 SYSTEMS THINKING

Imagine two people showing up in your neighborhood, after those orange traffic cones have been set out, digging up the street. You catch sight of a digging machine operated by one, and the other filling the ditch back up as soon as the first one moves. Wouldn't that be puzzling? You approach them and ask about the absurd job they're doing. They tell you that the other coworker who is responsible for laying the pipes in, had called in sick that day and they're doing their job without him.

What is wrong? Something big is wrong. They do lack the perception of process. If you asked the same question to each of these three coworkers that I just pictured for you in a scenario, you might be surprised to get three different answers. It could be one of the three scenarios: they have no vision of what they are doing beyond the task; they see beyond the task to a useful output, but not where this output fits in the big picture; they see themselves as part of a process and do have a vision of what they are trying to accomplish and that affects what they do. The more you look at the bigger picture, or are trained to do so, and the more connections you see, the better you'd be able to look at the process as a whole. Think about the cogwheels.

A process is a group of interrelated activities that use resources to convert inputs into outputs.

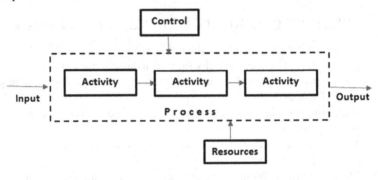

A Process

As a writer you should know that words both say and do things. Writing is the heart of your craft, an extension of the writer's identity. Silence, as they say, cannot be misquoted. The purpose of a piece of technical writing is to present information clearly and concisely so that it can be easily understood. Clarity therefore begins with the title if you desire to technically write right. The purpose of most work-related writing is to inform somebody of something, to persuade somebody to do something, to collaborate with someone on something, or any combination of these. Much like technical knowledge and skill, writing follows an orderly sequence of events that leads to clear and concise technical prose. The usual form for paragraphs in technical writings is to proceed from the general to the specific (deduction) rather than from the specific to the general (induction). Technical writing is expository writing which means it belongs to that great class of writing intended primarily to convey information to the user/reader. Its purpose is to get the message across quickly and clearly – no long paragraphs, complicated sentence structure, or fancy vocabulary.

Let's take a look at a few examples:

What follows here is a summary of those safety precautions in a short, clear DO and DON'T form. Read them. If you do not understand one of them, look back in the manual and read the what, where, and why of it. If you cannot find it in the manual, ask your supervisor and find out.

CAUTION

An operating procedure or practice that, if not correctly followed, could result in damage to, or destruction of equipment.

This has been extracted from a technical manual. Is the requirement correctly worded? It is a warning message that must have been left off from the review phase. Try to rephrase it.

Here's my proposed version:

If an operating procedure or practice is not correctly followed, serious damage to personnel or equipment may result.

One of the important lessons that I have learned as a human factors specialist is that our behavior is motivated by consequences. When we're told what an unsafe act is, immediately followed by why it should be avoided, then we become more conscious of the actions we take and activities we perform because we know about the probable consequences. It is the difference between understanding and knowing – the former playing a more important role in the human judgment, situational awareness, decision making, and problem-solving.

When you're writing and communicating to a tech audience, focusing on the why can be rewarding. Tell them what to do and what not to do when creating requirements, SOPs, guides, etc., but show them why and proactively disclose the truth. Unveil what might happen upon deviating from your prescriptions.

Example: Do not try to clean a large area all at once. Trying to clean a large area at once may allow solvent to evaporate, leaving the surface contaminated.

CAUTION

Make sure to seal or repair all damage. Failure to seal or repair damage can cause more internal damage because moisture can get in and freeze at altitude.

CAUTION

Remove all covers. Engine should not be operated with cover in place because the covers can come off and damage the engines.

CAUTION

When you use a towbarless tractor, make sure that you obey fully all the instructions in this procedure. If you do not, the tractor can cause important scraping or other damage to the nose landing gear and to the airframe structure around the NLG.

CAUTION

Do not attempt to remove foreign objects with air pressure (what/action). The objects may be lodged tighter or blown deeper into confined or inaccessible areas (why/ consequences).

Wind Warning

Do not jack aircraft on wing and fuselage jack points aboard ship if wind exceeds 15 knots and/or pitch, roll, or heel of ship is expected to exceed 3 degrees.

WARNING

Do not apply excessive force when resetting circuit breakers (what/action), as this may cause over-travel, leading to a premature failure and cause latent/undetected faults which can prevent the circuit breaker from functioning (why/consequences).

9.5   THOU, THE WORD MISER; THOU, THE WORD MASTER

Stick with simplicity in all tech writings that you do. Be a word miser (for brevity) by eliminating everything that doesn't add to your well-planned meaning. Rid your (written) product of long words, empty phrases, flabby language, and excess adjectives and phrases. Use active voice and present tense. Be a word master (for accuracy) and use words with respect and precision. Get specific. Mean what you say and say what you mean.

When writing is not clear, the thinking behind the writing may not be clear either. Decisions regarding a piece of tech writing are driven by two questions: Who will use it? What will they use it for? Not only should the writing be concise and quickly paced, it also needs to be flawless in language. To better pursue your goals as a tech writer, you should know who your readers are (their level of prior knowledge and the expectations they have while reading your document); use a consistent layout, format, and style; and write clearly and concisely with no ambiguity. The most critical sections are taking a series of instructions or steps – as concise as possible – in order to carry out a task. Concise isn't the opposite of long; it's the opposite of wordy. If information doesn't add value, simply leave it out. Often, the less you say, the more impact it has, provided that you are clear, concise, and valid. Bear in mind that your words will do rather than just say things – they prompt people to act upon reading.

9.6   CLOUDS THICK, WHEREABOUTS UNKNOWN

9.6.1   Relationships Transcend Parts

The goal of technical writing is to enable readers to use a technology or understand a process or concept. Technical writing style should use an objective, not a subjective, tone. Your writing style must be direct and emphasize on exactness and clarity rather than elegance or allusiveness. The delivery and success of technical communication also require that each author develop a style, an ability to use language and phrasing to assist rather than impede the reader's efforts to understand the information being communicated.

One important aspect of technical writing, I strongly believe, is the ability to view the material from the perspective of a typical user and be able to explain instructions clearly and simply. Who is the target? Your target audience. If a new hire and a supervisor are handed two copies of the same piece of technical writing – say, an SOP – they both should have the same understanding of that text.

Unexpected events and future surprises result from relationships and not parts. Several recent runway incursions, as statistics show, have been attributed to communications. According to the Federal Aviation Administration (FAA), the most important concept in pilot–controller communications is understanding. Pilots must acknowledge each radio communication with Air Traffic Control (ATC) by using the appropriate aircraft call sign and confirming all hold short instructions. But what rules, regulations, or requirements remain in force on the ground, on the individual's desks?

An important factor needed in the heat of battle is effective communication in which the coded message (sent) and the decoded message (received) are completely identical. Your selected method of delivery should fully match the target audience's literacy level and processing skills. Everything stems from conducting a thorough needs assessment to proactively determine all applicable requirements (regulatory and organizational) as well as the user's existing versus desired qualifications.

To make it syntactically, thematically, and contextually right, the user and the creator should interface effectively by providing feedback. If feedback does not complement the loop, then users will make excuses such as, "these tech pubs suck," to avoid using them.

9.7   IT IS WRITTEN; ARE THEY SOLD?

9.7.1   CATCH 22

When reading, you start a dialogue with the text – a dialogue that had earlier taken place between the writer and words. There's two-way communication established between you and the text that – if it is a set of instructions – is meant to tell you what to do and, more ideally, show you how to do it. Although your reader, the owner and user of your prescription, is in a position of extra-textual or outside the text, yet s/he is pretty much inside and part of it. The role of the reader in creation of meaning shall never be underestimated. In the act of delivering or transferring a shipment called "text" to a reader, the role of behind-the-scenes activities cannot be underestimated. It's not only the text that gets transferred, but also its building blocks, cultural attachments, and thought processes that come together to form the "text." The role of the reader in the creation of meaning completes the loop. Whatever is represented in a text will ultimately need to be remade, recreated, and remembered not necessarily as received, but as understood and used.

In this technological age, almost anyone with a responsible job writes and reads instructions. For example, you might instruct a new employee in activating his or her voice mail system or assist a customer in shipping radioactive waste or show them how to fix a flap asymmetry. The employee going on vacation writes instructions for the person filling in. Computer users routinely consult hard copy or online manuals (or documentation) for all sorts of tasks. Ask yourself the following questions when you find yourself in a position writing for other people:

- *Why am I doing this?*
- *How do I do it?*
- *What materials and equipment will I need?*
- *Where do I begin?*
- *What do I do next?*
- *What could go wrong?*

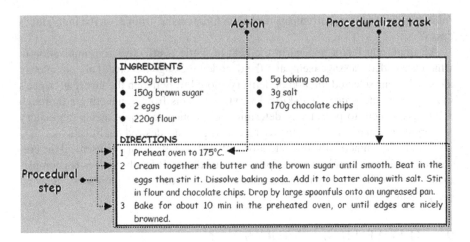

FIGURE 9.1　Components of a standard operating procedure.

To briefly summarize, look at an SOP that consists of a set of tasks that presents step-by-step instructions in the form of procedural steps composed of many actions (Figure 9.1).

Following a well-planned, well-developed standard operating procedure is easy and it can help in enhancing consistency of services. The roles of tribal knowledge or urban legend cannot be underestimated in using the tech pubs. Catch 22 is that we keep encouraging people to refer to and use the tech pubs as a fundamental principle, but we have made it a painful experience for them by, say, moving forward with not the best, but the lowest bidder. Instructions spell out the steps required for completing a task or series of tasks (say, installing printer software on your hard drive, operating an electron microscope, or taking an aircraft engine off for maintenance). The audience for a set of instructions might be someone who doesn't know how to perform the task or someone who wants to perform it by adhering fully to the documented requirements – I am one of those persons. In either case, effective instructions enable users to get the job done safely and efficiently.

Procedures, a special type of instruction, serve as official guidelines for people who typically are already familiar with a given task (say, evacuating a high-rise building). Procedures ensure that all members of a group (say, employers or employees) coordinate their activities in performing the task.

Because they focus squarely on the user – the person who is going to "read," "understand," "interpret," and then "act" – instructions must meet the highest standards to get the job done. In other words, instructions must be user centered, created for a human being to read and understand. If requirements are not clearly documented, then any expectation of full compliance would sound more like a joke.

9.8　EMPOWER THE USER

If requirements exist but are not documented, then we'd better call them a rumor (with a lot of room to err).

The breakdown in communication and a lack of situational awareness have been cited as key contributory factors in catastrophes and accidents.

Using the tech pubs is quite different from buying a new pair of shoes – doesn't leave you with that many options to choose from. If you fail to spark their interest and your design distances itself from being human centered, then, according to the users, you're signing yourself and your commodity up for the "tech pubs suck" movement. Using the tech pubs, on the other hand, would be quite similar to using a vending machine. I'm sure that you, like me, have come across vending machines that left you with confusion and frustration. To simply get some water, I spent 3 minutes on one of these UIs, while spending 5 seconds on another. Why? I – the user – as well as my expectations and analyses had been taken into consideration in the latter and completely left off in the former during the design and development phases. If your end users, along with their complicated cognitive characteristics, are not included in the design and development phases, then you're on the path of wasting many resources.

Tech writers are part of a team that must aim at empowering others to solve problems by giving them very clear steps to follow:

> "If your computer does not start, check the electrical plug."
> "When you receive this prompt on your screen, take this action."
> "When you receive this message on your monitor, go to ..."
> "If the test results are in the range of ..., do ..."

9.9 THE CUSTOMER CAN ALWAYS WRITE: CHOOSERS CALLED BEGGARS

Your phone rings and you pick it up. Suddenly there's a man or woman on the other end talking to you about making a donation, renewing a subscription, supporting a political candidate, or removing chemicals in your tap water. Sometimes they can go on for minutes before you have the opportunity to speak. Wouldn't you feel irritated and pissed? One-way communication bores. People wait in lines and buy tickets to see monologues performed on the stage – a one-man-show piece.

Successful organizations work hard at building a good communication infrastructure. The voice of customer (VOC), being an invaluable opportunity, is heard and acted upon by organizations who strive to thrive. These people make it easy for customers to communicate and voice their concerns. They hand out postage-paid feedback forms, set up a toll-free phone line, encourage customers to fill out online feedback forms, etc.

What is the benefit of that? It would enable you to correct problems before you develop a reputation for bad service. Word of mouth matters. Communication is key. Do not underestimate your audience and constantly think of their unique experiences. Build and maintain bridges, not walls.

10 Going with the Grain of the Brain

The human brain, among all other extraordinary things that it does, is a meaning-maker machine. It looks for patterns and, in the absence of one, the brain fills in the blanks and invents meaning/pattern. In other words, order is imposed upon chaos. If you leave blanks and question marks in your message, the brain will instinctively manage to fill them in with the most available – not necessarily correct – information.

As a writer or a speaker, do you tend to talk to or at your audience? What is your role in facilitating effective communication and the interpretation process? What your reader will gather is directly linked to how your writing is going to unfold.

10.1 EXPERIMENTAL WRITING VS. TECHNICAL WRITING

10.1.1 NONRECIPROCITY VS. RECIPROCITY

Fiction writers tell us stories to challenge our self-understanding, to arouse creativity in us, to ask us to produce, to complete, to always add, to fill in the gaps, to imagine, to contribute, and to forget about the real world and reside in the world they are creating parallel to the reality we have been accustomed to perceive and live with. Tech writers, on the other hand, tell stories to remove the challenge, to enforce obedience, to encourage clarity, to facilitate understanding and consistency, to request action, to halt innovation, to remove doubt, to be a good listener, to be a mere receiver and doer.

In the preface to his *The Spoils of Poynton*, Henry James writes, "Life being all inclusion and confusion, and art being all discrimination and selection" (James 1987).

In his "Decay of Lying," Oscar Wilde concludes and concurs that "Life imitates Art more than Art imitates Life" (Wilde 2010).

Four Kinds of Writing

NARRATIVE WRITING

The type of written language that is used to tell a story. Narrative text includes a starting point, events that typically lead up to a climax, then a resolution.

DESCRIPTIVE WRITING

The type of written language that is used to make a story come to life. Descriptive texts include vivid imagery, similes, metaphors or other figures of speech to explain how people feel, think or act.

PERSUASIVE WRITING

The type of written language that is used to make a point, take a side in a debate, and convince others of the validity of that viewpoint.

EXPOSITORY WRITING

The type of written language that is used to provide a detailed explanation of an idea, like a recipe, a set of instructions in an SOP, in a user-guide, etc.

Works of literature aim at hiding the truth, while technical writing unhides and clarifies it. The truth that a user would need to fully grasp, the artistic language would make imaginative and intuitive. Connotation overrides denotation. Imagination precedes and proceeds it. It is subjective rather than objective. A technical writer writes to express, not to impress. There would be no system for reconciling the opposites, as Samuel Taylor Coleridge has defined poetry.

Creative writing is freeing consciousness on the page: being inaccessible to the majority of readers and for being too exclusively concerned with the inner workings of an individual mind – style – rather than with the individual's relationship to the real world. In experimental writing there is more absence, more omission, both of which would be impossibly dangerous concepts in tech writing. However, you are likely to find a job in the technical field with a degree in English, creative writing, or a similar liberal arts subject if you focus on the technical aspects of those positions.

James Joyce once said, "The demand that I make of my reader is that he should devote his whole life to reading my works" (Ellmann 1982). As much as Joyce has been admired by readers and critics as one of the most important writers of the 20th century, just imagine if these words were coming from the mouth of a technical writer. Joyce's epic novel, *Ulysses* (1922) narrates the events of one day (June 16, 1904) in the lives of three Dubliners – Leopold Bloom, Molly Bloom, and Stephen Dedalus. Bloomsday is the name of this June day known to Joyceans around the globe, but if Joyce's technique – especially in the last 50 pages of Ulysses that describe Molly's flow of ideas, perceptions, sensations, and recollections without punctuation – were to be followed and practiced by tech writers, then it'd be the doomsday for their readers/users.

The text in front of you is a tool, an unstable medium. You have become part of it. This medium facilitates understanding. Your perception of the beautiful scenery outside depends on the quality of the windowpane. A professionally written text does function like a clear windowpane to help the reader see through the words and grasp the intended meaning. As a writer – fiction or nonfiction – you tell stories. As a nonfiction or tech writer, you would expect your reader (user) to buy in the truth exactly as you're selling it. While as a fiction writer, that buying in desire would be an illusion.

Language does not necessarily relate directly to objective reality. That's probably the main reason why symbols are being used to subjectively create a semi-objective reality among us. In masterpieces written by writers such as Samuel Beckett or Harold Pinter, language constantly fails its very primary mission: communication. Through repetition, these writers exhaust language to show how feeble words are. The more the words try (repetition), the more they fail to create a bond between the subjective world and the objective reality.

Bear in mind that:

- The same word can mean another thing in another context. Do not blind your reader by picking a vague perspective for them.
- Language is a system of differences. "Red" is understood not simply by relating this three-letter word to the color, but through the mediation of other colors.
- Words and language do not refer to things in the world, but to our concepts of things in the world.

An avant-garde poet and one of the founders of the Dada movement in the early 20th century, Tristan Tzara, wrote a poem in 1920, "To make a dadaist poem." It was much of a set of instructions:

TO MAKE A DADAIST POEM
Take a newspaper.
Take a pair of scissors.
Choose from this an article as long as you are planning to make your
  poem.
Cut out the article.
Next carefully cut out each of the words that make this article and put
  them all in a bag.
Shake gently.
Next take out each cutting one after the other.
Copy conscientiously in the order in which they left the bag.
The poem will resemble you.
And there you are – an infinitely original author of charming, even
  though unappreciated by the vulgar herd.
                                                      (D'haen and Bertens 2011)

Another good example is the OULIPO literary movement that formed in the 1960s in France by a group of writers and mathematicians. Jean Lescure's "N + 7" algorithm for transforming a text is an interesting example of the very movement. Take an excerpt from your favorite magazine, newspaper, novel, or holy book and replace each noun in it with the 7th noun following it in some standard dictionary. Let's have some fun by modifying the algorithm to replace every other noun, or take the 11th noun after the word in the dictionary. Would this still be called writing? Yes. But never practice it as a tech writer.

10.2 HERE IT IS, BUT YOU WON'T SEE IT: THE ART OF CONCEALMENT

Edgar Allan Poe was not a tech writer. Why? Because he was not talented enough, let's put it this way. Just kidding. Similar to other tales by Poe, "The Purloined Letter" is a tale of revenge. The story opens with the unnamed narrator and Dupin sitting in "a profound silence" in Dupin's "little back library." Their rumination is disturbed by the arrival of the prefect of police, Monsieur G —, who has come to consult Dupin for advice about a puzzling case. A letter containing valuable information is stolen by a state official. The utmost efforts of the police to recover the letter prove fruitless. They open all packages and books, examine chair legs with microscopes, remove all the carpet, and so on. But they cannot find the letter. Dupin, however, locates the letter by acting on the assumption that the thief would conceal it in an apparently visible place. By prominently displaying it on a pasteboard in the middle of his residence, the minister – who had stolen the letter – had created the appearance that the letter was simply part of the background rather than an obtruding figure.

The letter is constantly referred to, yet the content of the letter is never revealed to the reader.

Would a piece of tech writing be able to handle this? Absolutely not. Your art lies in unhiding and unveiling the information. One of the themes of this Poe story is that to hide something in plain sight just makes it extremely obvious. But how? If your writing does not show, then you must be preoccupied with the horrifying act of concealment.

10.3 SOFT: TO BE HARDLY NEGLECTED

> We spent over 50 years on the hardware, which is now pretty reliable. Now it's time to work with people.
>
> **– Donald Engen (former administrator of the FAA – 1986)**

I look at the fulfillment of requirements as a design – safety, quality, and compliance are at first puzzles that will be pieced together through team efforts in receiving, understanding, interpreting, and implementing the documented requirements. The most valuable resources and significant assets to get this sensitive task done are the true co-designers: the human assets (or the liveware).

Minimizing the occurrence of errors and eliminating the hurdles to communication can be ensured through high levels of staff competence, designing standard checklists, procedures, manuals, maps, charts, SOPS, etc. Mitigating risks and reducing the number of undesired consequences can be gained through cross-monitoring, cooperation, and effective communication.

Communicating requirements to the reader or user do not depend on the nature of the market. Human(s) write for humans to persuade them, to sell them an idea or ideas, to show them how to do things right, and how to do the right things.

When the task requirements exceed your resources' capabilities, then you're in big trouble.

Recent research has demonstrated that effective management of aspects of the messy day-to-day world is made possible through the deployment of non-technical skills. Non-technical skills are the cognitive and social skills that complement workers' technical skills. Our interpersonal skills could improve many things if only we were to put ourselves in someone else's shoes and view the problem, narrative, event, and solution through their eyes. I should make a bitter confession that in today's society, soft skills are hardly trained. In the aerospace industry, horizontal and vertical stabilizers are discussed more fully than the employees' personality stabilizers.

Errors occur when our actions deviate from our intentions. SOPs and tech writing determine actions. While different forms of error are associated with different types of mitigation, non-technical skills such as planning and preparation, monitoring and cross-checking, and evaluation of plans are important aspects of error detection and management. Undesired states are defined as the outcome of poor threat and error management (TEM). Developed by the late Professor Robert Helmreich and his colleagues at the University of Texas Human Factors Research Project, the TEM model describes that day-to-day operations do not always occur in a simple and benign environment in which the standard operating procedures as documented in the operational manual can be implemented. Rather, the context of day-to-day operations is messy and complex, and safe and efficient operations are maintained by operators effectively responding to and managing this complexity. In an elegantly simple way of describing this complexity, the TEM model categorizes three aspects of the messy day-to-day world that need to be the focus of specific actions to manage them: (1) threats; (2) errors; and (3) undesired states. Bad writing, being a potential threat, can increase the likelihood of errors on the user side.

During the 1960s and 1970s, advancements in technology resulted in dramatic improvements in aircraft reliability. As a result, human performance breakdown became much more evident in aircraft accidents because humans had to fit machines. Following a number of tragic aircraft accidents that plagued the aviation industry in the 1970s, the authorities took notice of the concept of crew resource management (CRM) training, whereby team members are taught to use all available resources: people, information, and equipment. The cockpit crew training may be considered a form of corrective action strategy because their performance is observed and corrected in a simulated environment to prepare them proactively. Concerned not so much with the technical knowledge and skills required to fly and operate an aircraft but rather with the cognitive and interpersonal skills needed to manage the flight within an organized aviation system, CRM aims at advocating assertiveness and teamwork by replacing the old-fashioned, one-way, authoritative Cockpit Run by Me methodology in a flaps up, gear up, and shut up dialogue from the pilot in command to the copilot. In 1996, authorities and researchers devised the NOTECHS (non-technical skills) checklist to measure the flying crew's soft skills: cooperation, leadership skills, situation awareness, and decision making.

```
1. Co-operation
- Team-building & maintaining
- Consideration of others
- Support of others
- Conflict resolving
```

```
3. Situation awareness
- Awareness of aircraft systems
- Awareness of external environment
- Awareness of time
```

```
2. Leadership & managerial skills
- Use of authority & assertiveness
- Providing & maintaining standards
- Planning & cooperation
- Workload management
```

```
4. Decision making
- Problem definition & diagnosis
- Option generation
- Risk assessment & option selection
- Outcome review
```

NOTECHS Checklist Items

10.4   IT'S IN YOUR HANDS

Our own relationship with language is unique and idiosyncratic. This relationship is formed by the books that we have read, by the people we have known, by the surroundings in which we have grown up, and by the languages we are familiar with. Language is not just a window on to a world, or a way of conveying information to the reader. It also has rhythmic, sensory, and visual aspects.

In her Nobel Prize acceptance speech in 1993, Toni Morrison told a story of a group of young people confronting an elderly blind woman, asking her to tell them whether the bird they held in one of their hands was dead or alive. The blind woman responded: "I don't know. I don't know whether the bird you are holding is dead or alive, but what I do know is that it is in your hands. It is in your hands" (Morrison 2020). By turning the tables on them, the blind woman's story, as Toni Morrison tells us, is also a metaphor for language. Words are always in your hands. Whether you create meaning out of them or you undermine the whole purpose. It's in your hands.

10.5   WRITING AND THE UNITY OF EFFECT

10.5.1   MODUS OPERANDI

In a good piece of writing, the whole does have a relationship to the parts. The whole does serve the parts, and the parts do serve the whole like a tree. Similar to a body, organs become the primary constituents whose absence of one would remove the concept of a complete body.

Edgar Allan Poe has left behind a unique literary legacy of poems, essays, and stories that have influenced writers and other artists throughout the decades. In his "On the Prose Tale," Poe makes mention of a very important concept in writing – a careful selection of the constituents (words) of your design (writing). Poe writes:

> A skillful literary artist has constructed a tale. If wise, he has not fashioned his thoughts to accommodate his incidents; but having conceived, with deliberate care, a certain unique or single effect to be wrought out, he then invents such incidents – he then combines such events as may best aid him in establishing this preconceived effect. If his very initial sentence tend not to the outbringing of this effect, then he has failed in his first step. In the whole composition there should be no word written, of which the tendency, direct or indirect, is not to the one pre-established design. And by such means, with such care and skill, a picture is at length painted which leaves in the mind of him who contemplates it with a kindred art, a sense of the fullest satisfaction.
>
> **(Poe 2006)**

Instead of didactic morality or personal self-expression, Poe was concerned about exploring the various and complex ways people form thoughts and ideas. In his 1846 essay, "The Philosophy of Composition," Poe underscores a more methodical and analytical approach. The starting point for his theory is that good writers should organize each of their creations (stories) around some already planned and intended response or "effect." A precise and focused method of composition should be followed to help a writer construct elements artistically enough to contribute to a "unity of effect." Outlining a general theory on the most effective way to construct a story, Poe says that the author purports to retail with perfect ease the "modus operandi" (the habit of working). By describing the writing process involved with one of his own poems, "The Raven," Poe shows how this theory works. He asserts that his poem was neither the product of "accident" nor "intuition." Instead, "the work proceeded, step by step, to its completion with the precision and rigid consequence of a mathematical problem."

Good writing is not carelessly created. It is a well-planned, carefully processed, and artistically created phenomenon.

10.6 FIVE ELEMENTS OF THE COMMUNICATION PROCESS

For effective communication to form and last, five elements need to exist. First, there must be a *sender*, someone who initiates the communication. The second required element is the *message*, which is the information the sender wants to convey. The third requirement for effective communication to take place is that there needs to be a *receiver*, or audience, for the message. This can be the *listener*, for a spoken message, or the *reader (user)*, for a written message. The sender needs to choose a way to transmit the message to the receiver. The choice for transmission is the *channel*, the fourth requirement for effective communication. A channel can take various forms. A speech, the telephone, or voicemail can be used to transmit spoken messages, while letters, memos, tech pubs, manuals, advisory circulars, airworthiness directives, service bulletins, e-mails, procedures, flowcharts, user guides, or fax machines can transmit written messages. The final required element needed for effective communication to occur is that there needs to be some *response*, or *feedback*, that the receiver/audience provides to the sender.

10.7   COMMUNICATION, THOU SHALT COMMUNICATE

Basic rules for effectively communicating include (1) ensuring that the message is clear, (2) verifying that the message is received and understood as intended by the sender, and (3) utilizing the appropriate medium for transmittal. To ensure the message is clear, the sender – the tech writer in our example – should:

- use appropriate terminology and avoid terms that are hard to understand;
- avoid lengthy explanations that dilute the point of the message;
- ensure the accuracy of the information; and
- adapt the level and approach of the communications to the intended audience.

11 Factors Called Humans

Machines are straightforward because they can work in predictable ways. Humans are infinitely more complex in the way they think and act, therefore it is much harder to analyze their behavior when, say, undesired circumstances arise. Among the definitions for safety out there, Dr. Scott Geller, a behavioral psychologist, provides one of the best versions, "Safety is a continuous fight with human nature. Because human nature (or natural motivating consequences) typically encourages at-risk behavior" (Geller 2001).

When deviance of any form becomes your preferred way of doing things, then normalization of deviance has prevailed. If humans are expected to comply with the documented requirements in order to avoid deviance proactively, then how those requirements are being presented to them and how well and proactively they've been planned would significantly count and vitally matter.

Human factors can be defined as a range of issues including the perceptual, physical, and mental capabilities of people and the interactions of individuals with other individuals as well as the equipment and the environment. As a discipline, human factors (HF) began in 1941 as an application of the principles of experimental psychology to the design of war equipment. It has expanded over the years into a distinctive discipline with many subareas. One of these areas – one crucial to HF – is the contribution of HF to design. Prior to World War II, the only test of the fit of the human to the machine was one of trial and error, in which the human either functioned with the machine (and was accepted) or could not (and was rejected). This Darwinian selection process would go on until a successful candidate was found, trained, and tasked accordingly.

Based on Heinrich's analysis of 75,000 industrial accidents, 88% of the accidents are caused by unsafe acts, 10% by unsafe conditions, and 2% by unpreventable causes.

Miscommunication of any form is an unsafe act that also creates an unsafe condition.

In highly regulated and sensitive contexts, requirements are written by blood. Not only do words say things, but they are also capable of doing things (Figure 11.1).

11.1 DESIGN ERRS: DESIGNERS OR MURDERERS?

Human factors issues related to flight-deck displays were recognized as early as 1923. An analysis of 19 airline accidents that occurred in the United States between 1991 and 2000 proves an attribution to crew error. The analysis indicated that in two-thirds of these accidents equipment failures or design flaws started the chain of events leading to the accident or inhibited recovery by making diagnosis of the

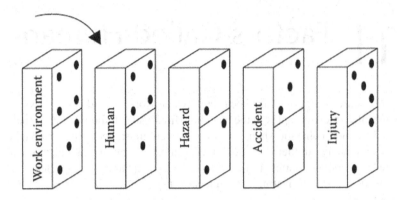

Classical accident model according to Heinrich.

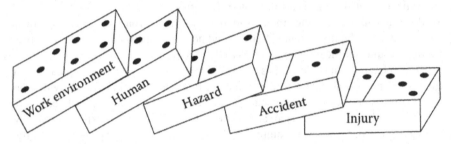

FIGURE 11.1   The domino effect.

problem difficult. Safety of any flight on this planet starts on the ground. Yes, designs do count.

11.2   CLOSE, YET DISTANT

With an average distance of 140 million miles (225 million kilometers) from Earth, curious humans are still able to communicate with Curiosity on Mars. "We're not sending robots to Mars. We're sending extensions of ourselves. These machines are us, and in visiting Mars, we may in time find that we are going home" (Manning and Simon 2010). An average of 14 minutes and 6 seconds will be required for signals to travel between Earth and Mars, meaning that if we could time travel, it'd take us less than 15 minutes to get to Mars from the third planet from the Sun.

Unlike the Mars Pathfinder, Spirit, and Opportunity innovative airbag-protected bounce and role landings in 1997 and 2004, the Mars Science Laboratory (MSL) represented the first use of a "soft landing" technique called the Sky Crane maneuver to land Curiosity on Mars in 2012. You tweak and tailor your approach and come up with a new strategy depending on the nature of the mission.

Rob Manning, the Mars Curiosity Rover Chief Engineer at NASA, shows us that Curiosity does have a shape. He writes,

> We're using a radical new landing design that has never been fully tested, because no tests on Earth can show how the design would fare on Mars. The needed actions are encoded in half a million lines of computer code, and there is zero margin of error.
>
> **(Manning and Simon 2010)**

Launching Voyagers into the space indicates our passion to make our presence in the universe known to the universe. We strive to communicate and make our voices heard.

However, we still find it challenging – if not impossible – to locate, understand, use, explain, comply, and communicate with the requirements sitting on a shelf 5 feet away from us, or presented to us as a software program on the computer screen facing us within 1 feet. Besides, the human (liveware–liveware) interaction, and acquisition of the required soft skills to best accomplish it, continues to present itself as a bottleneck in today's life – a world of independent, self-occupied individuals.

11.3 WHEN YOU WRITE, THEY ASSUME YOU'RE RIGHT

11.3.1 INSPECT WHAT YOU EXPECT

It is useful to review more general principles and methods for good technical writing. These principles are helpful in determining which unstructured outcomes to focus on in the report and how to present those outcomes: know your audience; define the objective(s) of the tech pub; keep in mind general objectives in your writing; construct an outline; use figures and tables; and use appendices to convey supplementary information.

The end user will have to spend time interacting with what you're creating with words, grammar, illustrations, figures, and punctuation – the ultimate design. If your writing is clear, but the layout or structure through which it's being delivered or presented is confusing, then you've failed your mission and the whole purpose of such creation would be undermined. Through writing, you try to help the user/receiver with solving a, say, puzzle. However, you might end up creating another puzzle for them and increase their workload if your creation deviates from your well-planned intentions or from the user's expectations.

Never allow your wisdom to get lost in your delivery. Lowest bidders do not necessarily provide the best quality.

11.4 THE STREET SMART, BOOK SMART TECH WRITER

That great writers are avid readers is a proven fact. People and places do shape our experience as writers and they get reflected in different forms – in our writings, too. After you spend some time in various countries, learning about their culture, myths, and legends, you'll be able to write about them. Hemingway's *A Farewell to Arms* draws heavily on the writer's personal experiences as an ambulance driver for the American Red Cross during World War I. Hemingway was book smart and street smart. Book smarts follow the rules and are good at doing it. How you would handle a tough situation is a world away from actually being in one. All teachers are former

students. A good tech writer needs to be attached to the requirements and tasks s/
he writes about.

11.5   UNDERSTANDING UNDERSTANDING

Once people learn the arbitrary symbols or grammatical rules, they are able to form
and then transfer mental representations to the mind of another person. We can speak
with a friend about our favorite foods, how we feel, what we're going to post on the
social media, and so on. In this context, based on the words we use and how they are
organized, the speaker and the listener will extract the correct or intended meaning.
However, language is powerful enough to betray your intentions.

In the works of the philosophical hermeneutics thinkers such as Friedrich
Schleiermacher and Wilheim Dilthey, understanding is defined as a process of psychological reconstruction of the mental experience. A full reconstruction of the mental
experience would be prerequisite to understanding. One big problem with language is
that it becomes limited by extra-linguistic experiences (Zen Buddhists, by encouraging the silencing of the talking mind, smuggle language into a state of lanuguageless
silence). Comprehensibility in communication implies that the message is received
and understood by the receiver exactly as intended by the sender. Research has shown
that most comprehensibility problems stem from a lexical and grammatical level.

Later on, as we see in the works of the poststructuralists and deconstructionists,
there are two levels of the text – the *declared* vs. the *demonstrated* – one being the
author's expressed intentions, whereas the other being the text's, along with all its
building blocks (words), actual representation or misrepresentation of intentions. We
all know of many examples in which the act of delivery deviates from the act of composition. Language is capable of betraying the composer's – the writer's or the speaker's
– intentions. Words are not owned by us. It's us who is defined and identified by words.

In section 32 of his phenomenal *Being and Time*, considered by many as probably the greatest book of philosophy written in the 20th century, Martin Heidegger
stresses that understanding must express itself in interpretation – the "as-structure"
of interpretation is the fulfillment of the "fore-structure" of understanding. Pre-
understanding precedes all understanding. As a writer, your understanding of the
subject matter you're writing about cannot be independent from your reader's understanding because, unlike experimental fiction, you're not using personal symbols or
metaphors to obscure your intentions. You are speaking directly to the reader (user),
not indirectly to them. You strive to be fully acknowledged and understood in your
tech documents. Misunderstanding would be fiction in your book.

In the not so pleasant "tech pubs suck" slogan, rather than understanding preceding interpretation, it is actually being preceded by the bitter interpretation in this
context. You wouldn't want to poison the water you're swimming in, would you?

Deceased scholar Dr. Richard E. Palmer (1933–2015), a mentor, a dear friend,
and a professor of literature and philosophy with whom I used to correspond for
about 14 years, shared with me his philosophy of understanding. To this decent,
generous scholar:

1. Understanding is an event.
2. An event takes place in a time and place.
3. An element that is often forgotten is that it takes place in a linguistic matrix, within a language.
4. So it is something that takes place in an existing individual, an individual with a past of experience, with this present, and with expectations of the future.

11.6 HUMAN INFORMATION PROCESSING

We may think in English, or in any other languages we're native in, but it won't sound odd to suspect that English is not the first language of English people. Thanks to science, we know that information processing goes on in the brains of babies even before they have learned to talk. It also occurs when people are actively thinking of different things and also when they are asleep.

The rise of the human information-processing approach in psychology is closely coupled with the growth of the fields of cognitive psychology, human factors, and human engineering. Differences in the cognitive environments of two people are simply differences in individual possession of facts, experience, and ability; people's stocks of assumptions vary according to their physical environments and cognitive abilities. The human mind has the capacity to manipulate the conceptual content of assumptions from a range of sources.

Never hesitate to resort to pictures, illustrations, or analogies when literal explanations fail to convince. Sometimes this is because an opposite simile can get a point across more quickly and easily. Become part of correct information processing by sticking with facts in your lucid narrative.

11.7 NONVERBAL COMMUNICATION

11.7.1 Economical Language Called Writing

Effective communication takes into account both the verbal and nonverbal aspects of communication. While verbal communication is organized by language, nonverbal communication is not.

You are driving to work early in the morning and it's still dark. Someone is driving toward you with their high beams on. You turn your high beams on to signal a message and the other driver immediately switches to low beams. Is this called communication with zero words being exchanged? Absolutely. You received the response you had asked for.

We have the capability of receiving information besides what is written or spoken. Our senses of touch, taste, sight, sound, smells, signs, symbols, colors, facial expressions, gestures, posture, and intuition are the primary sources of the nonverbal messages we receive. It is a silent language not formally taught, and which has existed long before language was invented.

With a shrug and a nod, an intention is communicated. Or one can read a sentence not only without moving one's lips, but also without speaking internally in thought. We all have the experience of words being on the tip of our tongues, or occasions when we know what we mean but can't remember or think of the right word.

The subject matter experts maintain that human communication is largely nonverbal. That is, although people communicate through words, the tone, rate, and volume of their speech, they also communicate through body language. Although body gestures play an important role in effective communication, tech writers are not blessed with this opportunity to be expressed visually in writing. There is no body language to aid understanding. In this case, writing tends to be more economical in its use of the language. There are no hesitators ("mmm," "er," "well," etc.) that litter our conversation. Written language is direct and efficient, as the writer would have to rely exclusively on language to communicate.

According to some studies, words are only 7% of the actual message we try to communicate. Tone of voice (voice inflection) accounts for 38% of our message, and nonverbal communication is by far the most important aspect, making up 55% of the actual meaning of the communication. Obviously, what counts is not *what* we say but *how* we say it. Words primarily address the content of our message. All three combined – words, tone of voice, and nonverbal communication – convey the emotional impact of the message.

A pat on the shoulder can be comforting when given by a good friend or a supervisor who is providing you with support, or it can be condescending when administered by a rival who says "better luck next time," after being promoted instead of you.

The meaning we assign to any communication is based on both the content of the verbal message and our interpretation of the nonverbal behavior that accompanies and surrounds the verbal message (Figure 11.2).

Stop

Both arms extended upward and crossed above head. During a pushback operation, the Stop signal will be used to tell the flight deck crew to apply brakes if the tow vehicle becomes disconnected from the aircraft.

FIGURE 11.2 Nonverbal communication, an example.

12 Enforcement, the Missing Link

If requirements are properly and adequately documented, and personnel are properly and adequately trained on using them, but the "enforcement" factor is absent from the compliance equation, then the organization will not be able to accomplish its mission and the whole purpose of having requirements documented gets undermined or will be limited to a check in the box to meet some regulatory requirement. Many standards by the regulatory agencies such as the FAA (Federal Aviation Administration), OSHA (Occupational Safety and Health Administration), and EASA (European Aviation Safety Agency) explicitly require the employer to train employees in the safety and health aspects of their jobs, stating that only employees who are "certified," "competent," or "qualified" are authorized to perform that specific task or job. It goes without saying that those who are behind creating such requirements are also required to meet certain criteria, especially when writing in English and having such a broad audience.

12.1 FROM THE AUTHORITY: FAA WRITING STANDARDS (ORDER 1000.36)

Good communication is fundamental to the safety and integrity of our airspace. Given that there are requirements for what to write but almost none for how to, or what not to, in a ten-page long document, the Federal Aviation Administration (FAA) has established writing standards for documents (FAA 2020). According to this Order (Figures 12.1–12.3),

> we must strive to communicate clearly with our customers and with each other. Over the years, our writing has become dense and needlessly complex, laden with technical terms and abbreviations that make it time-consuming and difficult for our readers to understand

03/31/03	1000.36

FAA Writing Standards

Chapter 1 General information

3. Why is FAA issuing this order? Since FAA's mission is so critical to both the safety and economy of our nation, we must strive to communicate clearly with our customers and with each other. Over the years, our writing has become dense and needlessly complex, laden with technical terms and abbreviations that make it time-consuming and difficult for our readers to understand.

FIGURE 12.1   The FAA document 1000.36.

h. Avoid using "shall." Shall is an ambiguous word. It can mean must, ought, or will. While shall cannot mean "should" or "may," writers have used it incorrectly for those terms and it has been read that way by the courts. Almost all legal writing experts agree that it's better to use "must" to impose requirements, including contractual requirements.

FIGURE 12.2   You shall not use shall.

Electronic Code of Federal Regulations

e-CFR data is current as of February 13, 2020

Title 14 → Chapter I → Subchapter C → **Part 25**

(j) *Flightcrew emergency exits.* For airplanes in which the proximity of passenger emergency exits to the flightcrew area does not offer a convenient and readily accessible means of evacuation of the flightcrew, and for all airplanes having a passenger seating capacity greater than 20, flightcrew exits shall be located in the flightcrew area. Such exits shall be of sufficient size and so located as to permit rapid evacuation by the crew. One exit shall be provided on each side of the airplane; or, alternatively, a top hatch shall be provided. Each exit must encompass an unobstructed rectangular opening of at least 19 by 20 inches unless satisfactory exit utility can be demonstrated by a typical crewmember.

FIGURE 12.3   Code of Federal Regulations (14 CFR Part 25) – Standard Airworthiness Certification Regulations.

12.2 WRITERS CALLED CULTURES

Those who are emotionally intelligent know what to say and when to say it. Real technical writers take this even one step further by knowing and practicing how to say what they want to say. Having zero interest in the divide-and-rule policy, this group does deconstruct the very policy by believing in the unite-and-pull policy. Addressing the explicitly stated needs and requirements, try to also go after the implied needs and requirements that the user is, due to any reason, reluctant or shy to talk about directly, but makes mention of them if you are able to read between the lines and listen carefully to what they don't directly communicate.

In their life cylce, documents are read, misread, interpreted, misinterpreted, misplaced, translated, mistranslated, mistreated, trained, complied with, deviated from, and the story goes on. Writing should solve a problem or, if proactively well designed, prevent it from happening. Not only does writing in a vacuum diverge and confuse, it also creates more problems for its audience. In order to help them solve a puzzle, you end up creating more puzzles for them, as discussed a few times in this book. Only after you have analyzed the need successfully and thoroughly with the user (co-creator), will you be able to accomplish convergence. A good way to address needs is to upgrade the *should* items to *shall* items.

My experience indicates that employees' perceptions of managers' commitment to the mission and vision of the organization is a major determinant of compliance and professionalism climate. The safety culture, for instance, is a long-term and systematic process consisting of two major components: the necessary framework within an organization (the management hierarchy) and the attitude of staff at all levels in responding to and benefitting from the framework.

Some lessons learned:

- Those who benefit from a broken system have no interest in fixing it.
- Independent cultures prefer minimized human interaction. Interdependent (or collective) cultures prefer otherwise.
- Teams go through a transformation – forming, storming, norming, performing.
- Overstating the obvious deems necessary, although it shouldn't.
- Those who are tasked with creating requirements must be aware that a car with the controls on the left-hand side is driven on the right-hand side of the road.

It's hard to change the culture and direction of a company. Maybe that's why entre-preneurs, having given up on changing the existing, rigid culture of other orga-nizations, create a new company and start pressuring the old, rigid culture of the company from outside. We must constantly bear in mind that once our beliefs form and get justified, it'll take more compelling evidence to change them than it did to create them.

12.3 INDEPENDENT AND INTERDEPENDENT

12.3.1 AN ENCOUNTER

Migration keeps the globe working. Like many of you, I have had law-abiding coworkers, friends, managers, and supervisors from all walks of life – lawful immi-grants and natives. This is making America more beautiful and more powerful. Organizations that deliberately take actions to be more diverse and inclusive would be able to identify their strengths and weaknesses faster and, ultimately, become stronger. Inclusion of more voices leads to provision of more input to the (upper) management to make better and more robust decisions. Listening to ideas contrary to the prevailing thought – dissenting opinions – is key because too much agreement is always dangerous and leads to the confirmation bias (we'll talk about this bias and other biases in the coming pages).

Those who have chosen to be in America, or as an immigrant anywhere else around the globe, hardly take things for granted and aim at maximizing their poten-tial. The precious experiences those individuals bring to the table along with their cultural values. This goes beyond the workplace and extends to families, communi-ties, etc. The interdependent or collective cultures have provided passionate human assets to the individuals' second countries of residence as immigrants.

Let me tell you more about my experience here. I grew up exposed to two worlds, two cultures – the traditional Eastern culture and the Western culture through travel-ing, studying their history, literature, and corresponding with scholars, writers, and thinkers. In my first culture of interdependence, he who teaches me a letter has made me his slave, as the society believed. In my second culture of independence, the power of "I will" kicked in.

> Now I go alone, my disciples. You too go now, alone. Thus I want it. Verily, I counsel you: go away from me and resist Zarathustra! And even better: be ashamed of him! Perhaps he deceived you. One repays a teacher badly if one always remains nothing but a pupil. You say you believe in Zarathustra? But what matters Zarathustra? You are my believers, but what matter all believers? You had not yet sought yourselves: and you found me. Thus do all believers; therefore all faith amounts to so little. Now I bid you lose me and find yourselves; and only when you have all denied me will I return to you.
>
> **(Nietzsche 1985)**

12.4   WHY REQUIREMENTS ARE NOT FOLLOWED

12.4.1   INHERITED THE MESS? WHY EMBRACE IT?

Among the effective ways used by organizations to close the gaps – after identifying them through conducting audits, inspections, or receiving – are training and proper documentation.

H. James Harrington suggests that people deviate from a prescribed process, procedure, or system because:

1. They misunderstand the procedures.
2. They don't know the procedures.
3. They don't understand why they should follow the procedures.
4. They find a better way of doing things.
5. The procedure as documented is too hard to do.
6. They don't have the knowledge or skills.
7. They were trained to do it differently.
8. Someone told them to do it differently.
9. They don't have the tools.
10. They don't have the time (Harrington 1991).

Each one of these reasons can be linked to either the "existence" or "lack" of something else – lack of oversight, lack of assertiveness, and lack of adequate training can be mentioned as three major root causes here.

13 Either Illuminate or Forever Eliminate

In one of Rumi's poems, we are told the story of some blind men in a dark room touching different parts of an elephant. One of them who is holding the tail announces that the elephant is thin and boney with stringy hair at the lower end. Another blind man who is touching the elephant's ear says, no, the elephant is a big, thin, triangular flap. Another one holding the trunk counters that, no, the elephant is a long muscular tube that blows air. No one is wrong; they each unknowingly have a partial answer to what is made available to them to perceive and judge (Figure 13.1).

FIGURE 13.1 One truth, many perceptions.

13.1 ONE TRUTH AND SO MANY REALITIES, JUDGMENTS, AND INTERPRETATIONS?

Similar to the very story is one of the most influential paradigms in Western thought exemplified by Plato's *Allegory of the Cave*. What the people in the cave take to be the real world is in fact only an illusion. They do not see the things as they are, but only the shadows of the things projected on the wall of the cave of the things that pass in front of the fire behind them. That's why Plato was not a fan of poetry or creative writing at all because of art misrepresenting the truth. The world of ideas (knowledge) versus the world of representation (perception) would also be an interesting topic for a tech writer to ponder upon. The more you lean toward talking about the knowledge or the concept, the more you help your reader with having the right perception. When you're writing down steps for, say, repairing a connector on the flight deck, departing from the subjective reality and arriving at the land of objective reality would tremendously help the user. Just be the writer you wish to be a reader of.

Our judgment is biased. Once you become aware of this, that awareness shows itself in your craft as a writer and allows for your wisdom to shine through the madness of words.

13.2 PATIENCE DOES PAY OFF

13.2.1 Excuse Me?

In his beautiful *Little Prince*, part of Antoine de Saint-Exupéry's narrative concerns the prince's encounter with a fox. When the prince asks the fox to come and play with him, the fox says he cannot play with the prince because he is not "tamed." Here's the rest of the story:

> "What must I do, to tame you?" asked the little prince.
> "You must be very patient," replied the fox. "First you will sit down at a little distance from me – like that – in the grass. I shall look at you out of the corner of my eye, and you will say nothing. Words are the source of misunderstandings."
> **(Exupéry 2020; Figure 13.2)**

If you don't agree with this World War II pilot, check out the following:

> We decided on the boat.
> I read the article on the train.
> We watched the birds over the bridge.
> She saw him with a telescope.
> Please listen to the music in the other room.
> They are hunting dogs.
> I heard the dog barking behind the house.
> The chicken is ready to eat.
> Visiting relatives can be fun.
> They are flying planes.

They saw a man on a horse with a wooden leg.

FIGURE 13.2 What did you say you saw?

TABLE 13.1
Language Development in Babies

	Language Development
Month (approximate)	Stage
4	Babbles many speech sounds ("bi-goo")
10	Babbling: resembles household language ("ma-ma")
12	One-word stage ("doggy")
24	Two-word stage ("Get ball")
24+	Rapid development into complete sentences

13.3   THINKING IN LANGUAGE: THE LANGUAGE-THOUGHT

Language is much more than a means of communication. It's also the raw material of thought. Language can be used both to illuminate and to obscure truth. Being wonderfully fluid, dynamic, and surprising, language can induce laughter, tears, anger, doubt, and pride.

Thoughts precede behavior and beliefs precede thoughts. Values, beliefs, awareness, and knowledge precede actions. How we have perceived the world does determine the way we also communicate and share our experiences. Which is truer to the facts?

Thought and language are closely – very very closely – interrelated. According to cognitive psychologist Lev Vygotsky (1896–1934), thought and language develop in a child independently till the age of 2. Prior to that age, thought is preverbal and is experienced more in action. Around 2 years, the child expresses his thought verbally and speech emerges as a tool for communication (Table 13.1).

13.4   CONTEXT IN THIS COURSE

After a month-long holiday in the United States, a couple finally boarded the plane in San Francisco on a Sunday heading home. As the plane reached cruising speed with the seat belt sign switched off, a gentleman with the build of Mike Tyson in the front row got up from his seat, turned to face the back, raised his arm and yelled, "HIJACK!"

Everyone was frozen to the seat, expecting the worst to happen. The two Air Marshals were about to jump on the guy and overpower him when another voice answered from the back of the plane: "HI JOHN!"

The moral of the story is: if you have a friend named Jack, for heaven's sake don't ever greet him on the plane. The context dictates our selection (of words, tone, behavior, etc.).

The social world is truly a world of communication, discourse, or rhetoric. Every text must be understood in a context and is approached with a context of questioning and prior understanding of the matter from the side of the reader. For instance, one must understand the language before one can make sense of a text in that language. One needs to know the historical context and likely purpose out of which a text arose. It is on the basis of this prior understanding that one construes a text. If a loan company sends you a letter with big letters on the envelope saying, "This is the last notice," will you become happy about it because you will receive none of those letters anymore? Words without contexts are just letters and signs (Figures 13.3 and 13.4)

Aoccdrnig to rscheearch at Cmabrigde uinervtisy, it deosn't mttaer waht oredr the ltteers in a wrod are, the olny iprmoetnt tihng is taht the frist and lsat ltteres are at the rghit pclae. The rset can be a tatol mses and you can sitll raed it wouthit a porbelm. Tihs is bcuseae we do not raed ervey lteter by it slef but the wrod as a wlohe.

FIGURE 13.3 We can read all or some of these texts without even thinking about them. Why? Because our brain tends to fill in gaps, imposes "order" upon "chaos" by expecting certain letters to be present.

Do yuo fnid tihs smilpe to raed? Bceuase of the phaonmneal pweor of the hmuan mnid, msot plepoe do.

FIGURE 13.4 We can read all or some of these texts without even thinking about them. Why? Because our brain tends to fill in gaps, imposes "order" upon "chaos" by expecting certain letters to be present.

The capacity to define the situation for self and others is a key dimension of social power. Language used in social, cultural, and historical contexts creates discourse which is a communicative event or a series of communicative events tied together by a common element such as a topic or point of view. A speaker creates discourse when talking, as a writer creates discourse when writing. If you write, you should be very careful with the discourse you're creating. Never unleash words in an uncontrolled, unmonitored manner.

You're not going to talk to an elderly man in a church the same way you are to a millennial in a bar. You're not going to talk to your mother over the Christmas turkey the same way you are to a coworker over a sandwich. You're not going to talk to an aircraft mechanic, in a tech manual, the way you talk to your favorite soccer player.

Know where you are and who you're with, and guide your words and actions accordingly.

Context plays an important role. If a stranger walks toward your house at the dead of night carrying a chainsaw, then you do have every right to feel apprehensive. But if a guy you have asked to cut down a tree walks toward your house with a chainsaw in the middle of the day, not only will you be pleased that he has arrived to get the job done, but you will also pay him for that. Technical writing is associated with a certain context that dictates rules and principles (Figure 13.5).

A statement or an argument contains a fallacy when it appears to be correct but on further examination is found to be incorrect. Aren't a tall toddler and a tall woman two very different things? Being a good soccer player doesn't necessarily make you a good husband, right? Refrain from comparing apples and oranges.

On February 23, 2010, it was reported that Democratic Senator Harry Reid had said that "if you're a man and out of work you may beat up your wife." From this, the reporters drew the conclusion that Reid thinks it is okay for men who are out of work to beat up their wives. In doing so, they committed a fallacy since the term *may* has more than one meaning. It can be used to mean "having the permission to" or express "possibility." In fact, Senator Reid had actually stated, "Men who are out of work tend to be abusive. Our domestic shelters are jammed." He was not giving unemployed guys permission to beat their wives.

Here is another example:

Using the term *right* as a moral right, whereas using it as a legal right are not the same. People might have had a legal right to own slaves – as did southern Americans before the Civil War – but (as all Americans would today concur) not a moral right.

We should use language and grammar properly so the meaning of our argument is clear. When in doubt of how to interpret a particular statement, we should ask the person to rephrase it more clearly. Rephrasing does not mean repeating the same thing on and on. Rewording plus the proper use of body language would make this transition smoother.

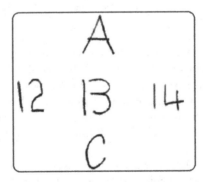

FIGURE 13.5 Interpreting the ambiguous characters that create the second items in both series is determined by the context.

Check out this dialogue from the 1996 movie *Spy Hard*:

Agent: Sir, we've intercepted a very disturbing satellite transmission from our listening post on the Rock of Gibraltar.
Director: What is it?
Agent: It's this really big rock sticking out of the water on the south coast of Spain.

In his *How to Do Things with Words*, J. L. Austin uses a sentence as an example, "France is hexagonal" (Austin 1962). Then he asks if this is true or false in terms of a sentence's mission being the reflection of reality in the world. His answer is that it depends. If you are a commander contemplating an impending battle and planning for it, stating that France is hexagonal might help you assess various military options of defense and attack. But if as a geographer you are tasked with mapping France's contours, then saying that France is hexagonal would not be sufficient and a greater degree of detail, study, and scale is required of a mapmaker. Therefore, as Austin concludes, "France is hexagonal" can be true and also false. Being true or false depends on a variable called context.

13.5 HUMANS WRITE, HUMANS READ, HUMANS WONDER, HUMANS GIVE UP

In his essay "Politics and the English Language," George Orwell writes,

> A scrupulous writer, in every sentence that he writes, will ask himself at least four questions, thus: 1. What am I trying to say? 2. What words will express it? 3. What image or idiom will make it clearer? 4. Is this image fresh enough to have an effect?
>
> **(Orwell 2000)**

Clarity is important in writing in general and in creating requirements by tech writers in particular. The average reader should not have any questions about the objective of the piece of writing. When writing, do sufficient research to assure you are compliant with all applicable requirements, as well as with the human understanding process. All stakeholders – the target audience included – should have an opportunity to have input and give feedback about the document before it gets officially released. Tech pubs play an important role in directing people on doing things right, rather than catching them doing them wrong.

Procedures tell us how to do something. They are the steps for getting things done. When documenting a procedure that stems directly from a requirement, we ask ourselves this question: What am I trying to accomplish? However, we are not alone in this process, as language – or the super powerful linguistic system – is preceding every single thought and action when doing so. A first offense against violators – who cut corners, take shortcuts, circumvent the process, or bypass the requirements – who deliberately choose to deviate from the prescribed implementation guidelines, may result in a one-day suspension; a second offense, a one-week suspension; and a third violation, termination. But, hold on for a second …

If requirements have been poorly documented, then who should be on the receiving end of such disciplinary action? Language? Probably. In most cases, language is the real culprit. Let's imprison it and rid ourselves of all existing and future misunderstandings. Bad documentation is the root of all evil. It doesn't matter how hard you're enforcing compliance with the documented requirements if those requirements are poorly designed and are not clear, concise, and valid (concise isn't the opposite of long; it's the opposite of wordy. A long report may be concise, while its abstract may be brief and concise); often, the less you say, the more impact it has.

Bear in mind that not all readers have good reading skills. English may be their second language. Remember code-switching? One of the rules in communicating effectively is the use of helpful language. This milestone may be achieved through identifying and avoiding its opposite: unhelpful language. Anything that blocks or prevents either party in a conversation – written or spoken – from understanding the other's point of view is called unhelpful language.

14 Storytellers That We Are
Art of the Narrative
That We Own

Humans are storytellers, just as writers are. We all have stories to tell as we live unique lives, but writing, being a challenge, draws us back. A story has a beginning, a middle, and an end, along with elements such as character, plot, setting, suspense, theme, and conflict. Besides pleasure, we tend to tell stories to fend something off. Shahrzad, the brave and smart narrator of the famous stories of *The Thousand and One Nights*, chooses to tell King Shahriar stories to postpone her death. In revenge for the infidelity of his wife, King Shahriyar has decided to marry a virgin every night and cut off her head in the morning. Shahrazad, Vazir's daughter, has also been sentenced to die in the morning. The king, who has been listening to Shahrzad's stories, wants to hear more stories. In this way, Shahrzad puts off the day of her execution until King Shahriyar relents. Shahrzad, a masterful narrator, draws the listener as well as the reader into a narrative whirlpool quite masterfully. Does she finish her story every night? She is too smart to do so. Technical writers – also storytellers – are no exception. They tell stories to fend things off: absence of clear instructions, lack of documented requirements, noncompliance, nonconformities, deviations, and violations.

There is an accident that involves two cars and two drivers. People gather and start to investigate and look for the driver at fault – as the judgmental human mind is accustomed to. By giving shape to the raw events, narrative is an account of events experienced by the narrator. When on a Monday morning you tell your coworker about the eventful weekend that you had, you are actually sharing a narrative with them. The narrative account of the events – say, endings – could easily get twisted should the narrator desire and choose so. By twisting the ending, a resentful antagonist becomes a lovable protagonist in the mind of the reader/listener.

Would that matter who was telling a story as long as there are still stories to be told and retold?

Narrative theory can be traced back to ancient Greece and Aristotle's *Poetics* written in 330 BCE. *Poetics* is a treatise on the making of a dramatic work of art in which Aristotle explains that drama is defined by its shape, composition, manner of construction, and purpose. In *Poetics*, Aristotle said that a narrative has a beginning, a middle, and an end. Ever since, scholars agree that sequence is necessary, if not sufficient, for narrative. The order of events moves in a linear way through time. A narrative is consistently responding to the famous question "and then what happened?"

In the second decade of the 20th century, it was the Russian Formalists who revolutionized linguistics and literary studies. They distinguished between the raw materials of the story – which they called "fabula" – and the means to convey them, the "syuzhet." Vladimir Propp made a substantial contribution when he wrote his groundbreaking study on the plot composition of the Russian fairytale. In *Morphology of the Folktale*, Propp put forward the radical idea that each of the plots in the hundred Russian fairytales that he had analyzed consisted of a sequence of 31 functions executed in an identical order. In his essay "Art as Technique" (1917), Viktor Shklovsky stresses the necessity of making things strange by "defamiliarizing" them in order to break with habitualization. According to this school of thought, art makes stone stony. While in literature (fiction), language aims at defamiliarizing, in technical documents (nonfiction), language pursues only one mission – that of familiarizing more and more.

In his *Poetics of Prose*, Tzvetan Todorov (1939–2017), the Bulgarian-French philosopher, has proposed a model of five stages in a narrative:

1. A state of equilibrium at the outset.
2. A disruption of the equilibrium by some action.
3. A recognition that there has been a disruption.
4. An attempt to repair the disruption.
5. A reinstatement of the initial equilibrium

(Todorov 1977)

14.1   LINEAR VS. NONLINEAR, THROUGHOUT THE YEAR

Stories with a beginning, middle, and an end follow a linear narrative structure. "Once upon a time" and "They lived happily hereafter" can begin and conclude a straightforward story and give the reader a good night kiss at the end. In a linear narrative, events unfold a step at a time in sequential real-time order. Mocking the whole idea of following a traditional structure and layout, the nonlinear narrative poses questions and evades answers. It leaves gaps behind, having no intention of filling them in. Writing technical documents, a set of instructions, a manual, an SOP, etc. follows a linear narrative with their readers (users) being the protagonist whose triumph would deem super easy. For the reader (user) to live happily hereafter, you – being the writer – should practice the lessons of victory where failure is not an option. "Tech pubs suck" would mark the moment of your death in this game.

One of the tested ways to ensure immortality in art and in writing is to keep readers trapped in visual, linguistic riddles. They find the language glass to be at best translucent, sometimes entirely opaque, and sometimes even a mirror reflecting back the image of the observer or reader. Georges Perec's *La Disparition* (*A Void*) is a novel of several hundred pages written entirely without the letter "e." In this book (a lipogram), Perec tackles the harrowing theme of loss and of the disappearance, evoking the loss of his parents during World War II and the disappearance of all those that, like his mother, were taken to the camps.

Let's explore some narratives together and ponder upon the shape of each for a few minutes:

"For Sale: Baby Shoes, Never Worn."
                    (A 6-word story attributed to Ernest Hemingway, Smith 2008)

Walter Benjamin – "The Task of the Translator":

In the appreciation of a work of art or an art form, consideration of the receiver never proves fruitful … No poem is intended for the reader, no picture for the beholder, no symphony for the listener.
                                                                      (Benjamin 1968)

Viktor Shklovsky – "Art as Technique":

And art exists that one may recover the sensation of life; it exists to make one feel things, to make the stone stony. The purpose of art is to impart the sensation of things as they are perceived and not as they are known. The technique of art is to make objects "unfamiliar," to make forms difficult, to increase the difficulty and length of perception because the process of perception is an aesthetic end in itself and must be prolonged.
                                                      (Julie Rivkin and Michael Ryan 2004)

Gustav Schwab – "The Horseman and Lake of Constance":

Schwab depicts a treacherous ride across a frozen lake. After braving this winter night, the Horseman arrives at a town across the lake and receives a warm welcome from its inhabitants. The townspeople gather to ask the horseman to recount his story of crossing the frozen lake, and it is only then that he realizes the danger of his journey when he had been completely unaware of the ice he was riding over. For no reason other than fear, the rider dies.
                                                                      (Schwab 2020)

James Joyce – *Finnegans Wake* (1939), a willfully obscure book:

What clashes here of wills gen wonts, oystrygods gaggin fishygods! Brékkek Kékkek Kékkek Kékkek! Kóax Kóax Kóax! Úalu Úalu Úalu! Quáouáuh! Where the Baddelaires partisans are still out to mathmaster Malachus Micgranes and the Verdons catapelting the camibalistics out of the Whoyteboyce of Hoodie Head.
                                                                      (Joyce 1939)

James Joyce – "The Dead":

LILY, the caretaker's daughter, was literally run off her feet. Hardly had she brought one gentleman into the little pantry behind the office on the ground floor and helped him off with his overcoat than the wheezy hall-door bell clanged again and she had to scamper along the bare hallway to let in another guest. It was well for her she had not to attend to the ladies also.
                                                                      (Joyce 1907)

Samuel Beckett – "Lessness":

RUINS TRUE REFUGE long last towards which so many false time out of mind. All sides endlessness earth sky as one no sound no stir. Grey face two pale blue little body heart beating only upright. Blacked out fallen open four walls over backwards true refuge issueless.

(Beckett 1995)

Samuel Beckett – *How It Is*:

alone murmur of millions and of three our journeys couples and abandons and the name we give to one another and give and give again

alone hear these scraps and murmur them in the mud to the mud my two companions as we have seen being on their way he who is coming towards me and he who is going from me something wrong there that is to say each in his part one

or in his part five or nine or thirteen so on

correct.

(Beckett 1964)

Samuel Beckett – *Molloy*:

All I know is what the words know, and the dead things, and that makes a handsome little sum, with a beginning, a middle and an end as in the well-built phrase and the long sonata of the dead. And truly it little matters what I say, this or that or any other thing. Saying is inventing. Wrong, very rightly wrong.

(Beckett 2012)

Roland Barthes – *S/Z*:

The goal of literary work (of literature as work) is to make the reader no longer a consumer, but a producer of the text. Our literature is characterized by the pitiless divorce which the literary institution maintains between the producer of the text and its user, between its owner and its customer, between its author and its reader.

(Barthes 1974)

Jacques Derrida – *Living on: Borderlines*:

A text lives only if it lives on, and it lives on only if it is at once translatable and untranslatable. … Totally translatable, it disappears as a text, as writing, as a body of language. Totally untranslatable, even within what is believed to be one language, it dies immediately.

(Derrida 1977)

Albert Camus – Nobel Acceptance Speech (December 10, 1957):

The artist forges himself to the others, midway between the beauty he cannot do without and the community he cannot tear himself away from. That is why true artists scorn nothing: they are obliged to understand rather than to judge.

(Camus 2020)

Vladimir Nabokov – "Essay on Metamorphosis":

Beauty plus pity-that is the closest we can get to a definition of art. Where there is beauty there is pity for the simple reason that beauty must die: beauty always dies, the manner dies with the matter, the world dies with the individual. If Kafka's "The Metamorphosis" strikes anyone as something more than an entomological fantasy, then I congratulate him on having joined the ranks of good and great readers.

(Nabokov 1980)

Ludwig Wittgenstein – *Tractatus Logico-Philosophicus*:

The world is my world: this is manifest in the fact that the limits of language (of that language which alone I understand) mean the limits of my world.

(Wittgenstein 2001)

James Joyce – "A Little Cloud":

Little Chandler felt his cheeks suffused with shame and he stood back out of the lamplight. He listened while the paroxysm of the child's sobbing grew less and less; and tears of remorse started to his eyes.

(Joyce 1907)

Umberto Eco – *The Name of the Rose*:

Perhaps the mission of those who love mankind is to make people laugh at the truth, *to make truth laugh*, because the only truth lies in learning to free ourselves from insane passion for the truth.

(Eco 1986)

George Orwell – *1984*:

On each landing, opposite the lift-shaft, the poster with the enormous face gazed from the wall. It was one of those pictures which are so contrived that the eyes follow you about when you move. BIG BROTHER IS WATCHING YOU, the caption beneath it ran.

(Orwell 2002)

Thomas Hardy – *The Mayor of Casterbridge*:

ONE evening of late summer, before the nineteenth century had reached one-third of its span, a young man and woman, the latter carrying a child, were approaching the large village of Weydon-Priors, in Upper Wessex, on foot. They were plainly but not ill clad, though the thick hoar of dust which had accumulated on their shoes and garments from an obviously long journey lent a disadvantageous shabbiness to their appearance just now.

(Hardy 1998)

E. E. Cummings, *Complete Poems*:

> where's Jack Was
> General Was
> the hero of the Battle of Because
> he's squatting
> in the middle of remember
> with his rotten old forgotten
> full of why
> (rub-her-bub)
> bub?
> (bubs).

(cummings 1991)

Aircraft Angle of Attack (AOA) Sensor – Removal Procedure:

Safety: Never hold the sensor by the vane. You can cause damage to yourself and to the sensor.

Look around carefully and do the job safely.

In the sensor you're working on, remove the potting compound from the screw.

There are 7 screws. Loosen them.

Now remove the blanking plate.

Remove the screw.

Now, carefully remove the plate. Do not cause damage to the sensor.

Carefully detach and hold the sensor.

Disconnect the connector.

Remove the sensor.

Remove the seal and discard it. (Figures 14.1–14.4)

PUSHBACK

A. Inspect the towbar to ensure it is in good operating condition.

B. Install the steering bypass pin into the nose gear steering bypass system. The bypass pin receptacle is located on the front side of the nose gear assembly. Move the bypass lever to the right and insert the bypass pin fully into the receptacle.

C. Connect the towbar to the aircraft; this should always be accomplished by two persons. The retractable pins on the towbar must mate with the nose gear attachment lugs. Once both pins have mated properly with the attachment lugs, insert the safety pin.

D. Connect the pushback tractor to the towbar. The towbar must be in line with the aircraft to ensure proper connection to the towbar and to ensure initiation of the pushback will be straight down the lead–in line.

E. Commence pushback operations as outlined in the manual.

FIGURE 14.1   A procedure for an aircraft pushback.

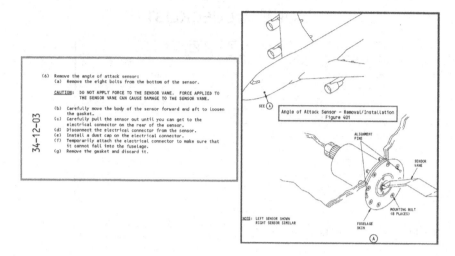

FIGURE 14.2 Angle of attack sensor location and removal.

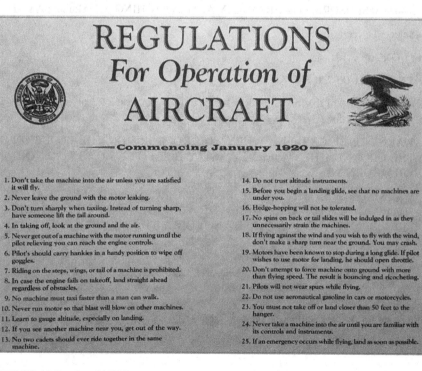

FIGURE 14.3 An old SOP.

NORMAL CHECKLIST

BEFORE TAKEOFF	
FLIGHT CONTROLS	CHECKED (BOTH)
FLT INST	CHECKED (BOTH)
BRIEFING	CONFIRMING
FLAP SETTING	CONF____(BOTH)
V1. VR. V1/FLEX TEMP....................	____(BOTH)
BUMP/DE-RATE	AS RQRD
ATC+TCAS	SET
RADAR + PWS	SECURED FOR TO
CABIN	TO NO BLUE
ECAM MEMO	

FIGURE 14.4 Aircraft taxi, a checklist.

EXAMPLE OF AN SOP IN AN AIRCRAFT MAINTENANCE MANUAL

34-11-00 SENSORS, POWER SUPPLY AND SWITCHING   Manual: AMM

Selected effectivity: ALL

** ON A/C ALL

TASK 34-11-00-720-802-A

Functional Test of the Insulation Resistance of the AOA Probe Resolvers from the Aircraft Inner Side

WARNING: BE CAREFUL WHEN YOU DO WORK ON THE PROBE ON THE GROUND. THE PROBES CAN STAY HOT FOR A LONG TIME AND WILL BURN YOU.

WARNING: OPEN THE PROBE HEATER CIRCUIT-BREAKERS 1DA1(2), 3DA1(2, 3) AND 5DA1(2, 3) BEFORE YOU OPEN THE PHC CIRCUIT BREAKERS 2DA1(2, 3). IF YOU DO NOT OBEY THIS SEQUENCE:

- THE PROBES WILL BECOME TOO HOT.
- THIS CAN CAUSE DAMAGE TO EQUIPMENT AND INJURY TO MAINTENANCE PERSONNEL.

Reason for the Job: Self explanatory. (Table 14.1)

TABLE 14.1

Safety Risk Mitigation Approaches – International Civil Aviation Organization (ICAO)

ICAO – Doc9859 (Safety Management Manual)
The three generic safety risk mitigation approaches include:

a) Avoidance. The activity is suspended either because the associated safety risks are intolerable or deemed unacceptable vis-à-vis the associated benefits.

b) Reduction. Some safety risk exposure is accepted, although the severity or probability associated with the risks are lessened, possibly by measures that mitigate the related consequences.

c) Segregation of exposure. Action is taken to isolate the potential consequences related to the hazard or to establish multiple layers of defenses to protect against them.

14.2 NARRATIVES AND THEIR INTERTEXTS

14.2.1 ORIGINALITY: A FALSE MENTALITY

Intertextuality tells us about relationships that link texts to other texts or parts of the same text. The text's dependence on prior words, codes, concepts, conventions, connotations, and, of course, other texts is an undeniable fact. How could you understand an SOP without linking it to your prior knowledge, understanding, and judgement? Being an intertext, every text borrows, knowingly or not, from the immense archive of previous culture.

In "Pierre Menard, Author of the Quixote," Mr. Borges describes a project by a 20th-century fictional author who is rewriting Miguel de Cervantes' Don Quixote. Pierre Menard produces one identical paragraph of the original text, a phenomenon achieved through multiple revisions until the words become Menard's own as well as the original writer's (Cervantes). Ultimately, the original 17th-century text – Don Quixote – becomes a product of the 20th-century in this process by Pierre Menard. This final product by Pierre Menard is seen by Borges as a "palimpsest" where traces from previous writing become clearly visible. And, ultimately, it is the reader who rewrites the text and takes full ownership of it in order to follow it. To own, you need to recreate. To recreate, you need to establish ties with that creation. Words acquire meaning through representing thoughts. The origin of all thought is sense experience.

One of the main types of intertextuality, as discussed by the French theorist Gérard Genette, is paratext – prefaces, epigraphs, epilogues, titles, forewords, footnotes, acknowledgments, allusions, references to other works, echoes, quotes, or citations.

Architextuality, another example for intertextuality, is a helpful tool in a writer's toolbox. It is the relationship of a text to other texts in the same category.

For example, the connection and similarities between technical manuals of an Airbus A320 Family with those of a CRJ-200, Embraer140 (regional jets), C-130 or an F/A-18.

14.3   BE FOR THE LAW

Language is powerfully capable of asserting and denying facts. We cannot hope to understand the role of language in communication if we allow the limits of language to exclude us from our role and use it as an excuse to absolve ourselves of our mission and responsibilities. As a technical writer, you'd be doing it best if you let your reader's (user's) voice speak through your voice as the narrator. This is doable through gaining some knowledge in the human factors, cognitive science, and psychology. It can be done.

As a technical writer, never make the process sound like a journey from innocence into experience. If you don't know how to mean what you mean, then you should probably refrain from writing. As much as you hate to be misunderstood, you won't be able to remain in control of the consequences should the words and language decide so on your delivery.

Kafka's relatively short story which is not much more than a page long, "Before the Law," is a parable about access and denial in the context of the law. We see a man from the country seeking to be admitted to the law. The doorkeeper, however, standing before the law, insists that the man cannot be granted admittance. Becoming a victim of the chaos before the order, the man from the country ends up spending his whole lifetime at the gate of the law. The doorkeeper's intimidating language prohibits the man from making the next move. The man doesn't even bother to challenge him, though.

Growing old and increasingly blind, the man approaches death. When he asks the doorkeeper why no one else attempted to gain admittance through the door, the doorkeeper responds, "No one else could ever be admitted here, since this gate was made only for you. I am now going to shut it" (Kafka 1971).

As we see, the door remains open, but the doorkeeper keeps excluding him from the equation, from his freedom. For you as a tech writer there is one law: to grant immediate admittance to the reader who is seeking the truth in your written work and not keeping them at the threshold of hesitations. Your writing is similar to that door, unique for each individual, to connect them with the intended, planned objective through law, order, and method.

Before the law stands a doorkeeper. To this doorkeeper, there comes a man from the country and prays for admittance to the law. But the doorkeeper says that he cannot grant admittance at the moment.

Allow me to rewrite this passage:

> Before understanding and compliance stands the tech writer. To this tech writer there comes an apprentice from a shop and prays for admittance to the world of the tech pubs. The tech writer immediately grants admittance to the apprentice, with no reservation, without having them wait. They live happily hereafter by continuing to communicate effectively.

14.4 COMPREHENSIBLE – DESIRE VS. AVOIDANCE

Communication, a skill central to the human experience, is a creative process which involves an unbounded and unknowable number of factors and viewpoints. Communication is more than sending and receiving messages; if the message has been delivered but not understood, no communication has taken place. Although we think we know how to do it and have done it since birth, as tech writers we have transitioned from an individual and independent user of language to a more group and interdependent user of this significant tool.

One key to successful communication, then, is to share an adequate amount of information in a skillful manner. Communication is the tool that lets us tell the hair stylist to take just a little off the sides, direct the doctor to where it hurts, inform the plumber that the broken pipe needs attention now, and tell the aircraft mechanic how to troubleshot an angle of attack sensor.

Comprehensibility means that the message is understood by the receiver in the way in which the sender intended, and research has shown that most comprehensibility problems occur at a lexical and grammatical level.

14.5 RECEIVED AS SENT? NEGATIVE, RESEND

Communication is sharing of meaning through the transmission of information via mutually understood signs. The quality and effectiveness of communication is determined by its intelligibility: the degree to which the intended message is understood by the receiver.

As a rule, communication signifies a sender (speaker, author, artist, situation, etc.) communicating to a receiver (listener, reader, spectator, observer, etc.). In order to make communication possible, the sender has to translate his thoughts or his knowledge into recognizable signals. He needs to encode them. The receiver, on the other hand, will have to understand and interpret those signals. He must decode. Grasping meaning depends on previous experiences, expectations, and apprehensions. This way, some messages may be misunderstood, some not understood at all, although encoded, transmitted, and received.

Sender (Encodes) → Message → Receiver (Decodes) → Receiver Becomes Sender and Encodes → Message → Receiver (Decodes)

Some of the factors affecting how a message is received include:

1. The receiver understanding what s/he chooses to understand.
2. The sender and receiver having different perceptions.
3. The receiver evaluating the message before accepting it.
4. Words meaning different things to different people in different contexts.

An important goal of a communication strategy is to convey information in as many different methods as possible. Peter Drucker, management guru and author (1909–2005), once said, "The most important thing in communication is hearing what isn't said."

15 Fully Owned by Language

We cannot separate the way we communicate with who we are. Understanding our own style and those of others would facilitate smooth communication.

Reality and perception differ as entries in the dictionary, but telling them apart wouldn't be easy because the same perception becomes the reality we seek. We explore a short story, a novel, a play, an article in a newspaper, a standard operating procedure, or a process and interpret them based on their connection with the linguistic part of our brain. Language defines us, directs us, cheers us up, disappoints us, talks about us, moves us, makes us cry, makes us laugh, makes us wonder, and makes us talk. Language helps, confuses, promises, fulfills, postpones, delays, builds, destroys, introduces, concludes, expands, contracts, discloses, hides, loves, and hates. Language becomes us, forms our identity, introduces the universe to us, and introduces us to the universe. Language always survives. Being the problem, language is the only solution as well.

15.1 PERCEPTION: MISCONCEPTIONS, MISJUDGMENTS, MISUNDERSTANDING, MISINTERPRETATION

A heuristic is any rule of thumb, a simple thinking strategy that often allows us to make judgments and solve problems speedier, but also more error prone. This does not always give us the right answer, but it is a helpful aid. Daniel Kahneman and Amos Tversky began investigating how intuition works in 1969. Their experiments showed that in complex situations, intuition doesn't work very well, but our fast-thinking minds make judgments anyway, introducing biases into our decisions.

When people have the opportunity to collect information from the world, they are more likely to select information that supports their preexisting attitudes, beliefs, and actions. People tend to have biases that they apply when making attributions. When evaluating others, many people underestimate the influence of external factors and overestimate the influence of internal factors. Many thinking biases are rooted in our evolutionary history; some are rooted in cultural traditions; and some are due to a variety of personal and social factors. Technical writers, whose work is taken very seriously, are not immune, either.

Tech writing means decision making – choosing between what to use and what to leave out.

15.2   BIASES: PROFOUNDLY AVAILABLE

We make hundreds of decisions and judgments every day. We admire, compliment, criticize, and complain. I personally privilege "that's a great idea" or "that'd be a good point of departure" over "that's a great manual." The availability bias involves making quick judgments based on the speed with which memories are aroused and become available to the conscious mind. According to the availability bias, we create a picture of the world using the examples that most easily come to mind. It's the comparison between what we really know compared to what we think we know. As the German philosopher, Arthur Schopenhauer, observed in 1851, "Every man takes the limits of his own field of vision for the limits of the world" (Schopenhauer 2013).

Perceptions follow beliefs, rather than vice versa. The availability bias is the tendency to assume that if we easily think of examples of a category, then that category must be common. However, this heuristic leads us astray when uncommon events are highly memorable. Assuming that other people perceive and interpret things the same way we do is called false consensus bias. We fall prey to the very bias when we project the way we think onto everyone else and assume that everyone else must certainly agree with us. Being a technical writer, you must constantly reevaluate your judgment to make sure it's refined from such destructive forces. Unbiased writing requires well-planned teamwork among open-minded individuals (Figure 15.1).

Noticing and remembering information and events that support the beliefs we already have about something, while simultaneously failing to notice or remember information or events to the contrary, are called confirmation bias. Psychologist Carol Tavris uses a good example by reversing the common belief of "seeing is believing" with "believing is seeing." Our beliefs and judgment – clouded by biases and misconceptions – form our reality. We perceive the world through our preexisting values that constantly impose order upon the perceived chaos. As a matter of fact, writers do fall victim to this bias a lot when their writing gets subconsciously wrapped around their core beliefs and create a totally detached perspective than that of their reader's. Cognitive conservatism is seen as a close cousin of confirmation bias. It refers to the humans' natural reluctance to admit mistakes or update beliefs.

Cognitive biases do not have a short list, and they must be considered as a hazard and treated proactively by looking for disconfirming evidence. Independence of judgment can hardly be declared.

FIGURE 15.1   Email from an engineer friend who has no biases.

15.3   IMMIGRANTS ARE SHY – DIGITAL LITERACY

Computers are stressors. Reading on paper can be a more pleasant experience for some people as compared to reading on screen because of their lack of computer literacy to effectively communicate with the UI. During this significant and quick transition, you don't want to lose these companions, some of whom used to be big heads before retirement. Technological advancements are taking place so rapidly that you could turn into an immigrant with all the culture shock they would be up against. Removing obstacles to technology acceptance must be a top priority through training and, most importantly, by creating more user-centered UI designs that would solve problems instead of adding to them.

As much as we all hate to lose our memory, there have been people in history who amazed others by showing off extraordinary talents. In *Brain Longevity: The Breakthrough Medical Program That Improves Your Mind and Memory*, we read:

> In the last years of his life, the philosopher William James became concerned about his memory. So, in a powerful display of will, he sat down and memorized all of John Milton's twelve-volume Paradise Lost. It took one month, and was, he said a "supremely inspirational" experience. But you do not have to be a "great thinker" to develop an extraordinary memory. Napoleon could greet many thousands of his soldiers by name. The businessman Charles Schwab knew the names of all eight thousand of his employees. The politician James Farley was able to call about fifty thousand people by their first names. General George Marshal was able to recall virtually every single event of World War II. The conductor Arturo Toscanini could remember every note of every instrument for one hundred operas and two hundred and fifty symphonies. (Khalsa 1999)

16 Maintenance – Continuing Airworthiness

According to Title 14 of the Code of Federal Regulations (CFR – Chapter I, Subchapter A, Part 1, §1.1), "Aircraft means a device that is used or intended to be used for flight in the air." The modern military aircraft is a collection of interdependent subsystems designed for a specific role. The modern civil aircraft is similarly a collection of subsystems identical to the military aircraft despite some significant differences.

Maintenance jobs typically start and finish with documentation. Documents not only convey instructions about task performance, but they also play an important part in communication by recording the completion of tasks and the extent of system disturbance. A study of the normal day-to-day activities of airline maintenance personnel found that for much of the time they were not touching aircraft at all, but were using technical logs, task cards, and maintenance manuals or were signing off tasks. The more unfamiliar the task, the more time was spent dealing with documentation.

Maintenance, defined by ICAO (International Civil Aviation Organization) is the performance of tasks required to ensure the continuing airworthiness of an aircraft, including any one or combination of overhaul, inspection, replacement, defect rectification, and the embodiment of a modification or repair.

Continuing airworthiness is defined by ICAO as "The set of processes by which an aircraft, engine, propeller or part complies with the applicable airworthiness requirements and remains in a condition for safe operation throughout its operating life."

The Federal Aviation Administration (FAA) and the European Aviation Safety Agency (EASA) require the Aircraft Type Certificate holder to prepare and revise the initial minimum scheduled maintenance requirements that are applicable to a dedicated aircraft (Regulatory Requirement CS/FAR 25.1529). This document is called the Maintenance Review Board Report (MRBR) and provides the scheduled maintenance tasks and their frequencies (intervals) for the aircraft systems (including power plant), structure, and zones.

Given the importance of paper records in maintenance, it is not surprising that poorly designed documents lie at the heart of many incidents. Procedures that are ambiguous, wordy, or repetitive are likely to promote errors. Procedures that are unworkable or unrealistic are likely to promote violations. While rewriting an organization's documentation may not be a feasible short-term objective, some improvements can be made incrementally. Simplified English, for example, can make the language of maintenance documentation clearer and more accessible, particularly in the case of staff for whom English is a second language. Even small improvements in page layout, diagrams, warnings, and illustrations can help to reduce errors.

For example, many companies print maintenance documentation in upper case, even though it has been known for many years that such text is more difficult to read than words written in the usual mixture of lower and upper case. I have been told by my English teachers that only abbreviations and acronyms may appear as uppercase. Replacing blocks of upper case text with normal mixed-case text can increase reading speed by 14% (Figure 16.1).

16.1   REGULATION, NOTICE, GUIDANCE

Airworthiness directives (AD): these are legally enforceable regulations issued by the authority making particular mandatory actions (changes, inspections, etc.). The FAA issues the ADs in accordance with 14 CFR part 39 to correct an unsafe condition in a product (an aircraft, engine, propeller, or appliance). ADs are mandatory airworthiness actions directed by the exporting authority to correct unsafe conditions experienced during operation before application for the FAA approval (Figures 16.2 and 16.3).

Service bulletin (SB): it is a notice to an aircraft operator from a manufacturer to inform the operator of a product improvement. An SB is issued when an unsafe

Technical Documents Review Sheet				

TDR No.	1		Effective Date:	27.07.2015
Revision			Revision Date:	27.07.2015
Subject	Activate all chemical oxygen generators in the lavatories until the generator oxygen supply is expended			

Document Reference Tree	**Typ**	**Document No.**	**Rev.**	**Title**
	AD	2012-11-09	1	Chemical Oxygen Generators in the lavatory

Applicability	Component (description in notes)	Notes: (Reason, configuration, model, description etc.)
	☒ Applicable (see notes) ☐ Not applicable (see notes)	Chemical Oxygen Generators

Compliance	☒ Mandatory	☒ Safety related	☐ Reliability improvement
	☐ Non mandatory	☐ Non safety related	☐ Design improvement
	☐ Optional	☐ Desirable	☐ Operations improvement
	☐ Recommended	☐ Maintenance Schedule	☐ Environmental improvement
	☐ Not applicable	☐ Other	☐ Customer request

FIGURE 16.1   Example of a technical document with action items.

FAA
Aviation Safety

AIRWORTHINESS DIRECTIVE

www.faa.gov/aircraft/safety/alerts/
www.gpoaccess.gov/fr/advanced.html

2012-11-09 R1 Transport Category Airplanes: Amendment 39-18221; Docket No. FAA-2015-2962; Directorate Identifier 2015-NM-071-AD.

(a) Effective Date

This AD is effective July 27, 2015.

(b) Affected ADs

This AD revises AD 2012-11-09, Amendment 39-17072 (77 FR 38000, June 26, 2012).

(c) Applicability

This AD applies to transport category airplanes, in passenger-carrying operations, as specified in paragraph (c)(1) or (c)(2) of this AD.

(1) Airplanes that complied with the requirements of AD 2011-04-09, Amendment 39-16630 (76 FR 12556, March 8, 2011).

(2) Airplanes equipped with any chemical oxygen generator installed in any lavatory and are:

(i) Operating under part 121 of the Federal Aviation Regulations (14 CFR part 121); or

(ii) U.S. registered and operating under part 129 of the Federal Aviation Regulations (14 CFR part 129), with a maximum passenger capacity of 20 or greater.

(d) Subject

Air Transport Association (ATA) of America Code 35, Oxygen.

(e) Unsafe Condition

This AD was prompted by the determination that the current design of chemical oxygen generators presents a hazard that could jeopardize flight safety and the discovery that certain existing requirements could impose an unnecessary burden on operators. We are issuing this AD to eliminate a hazard that could jeopardize flight safety, and to ensure that all lavatories have a supplemental oxygen supply.

(f) Compliance

Comply with this AD within the compliance times specified, unless already done.

FIGURE 16.2 An airworthiness directive issued by the FAA.

condition shows up that the manufacturer believes to be a safety hazard. Service bulletins may also result in airworthiness directives to be issued by the regulatory agency (FAA) (Figure 16.4).

Advisory circulars (ACs): to provide guidance for compliance with airworthiness regulations, the FAA issues the advisory circulars. Because a regulation could be interpreted in different ways, an AC can offer specific guidelines and give a standardized interpretation particularly when the regulations or requirements are otherwise vague (Figure 16.5).

EASA	AIRWORTHINESS DIRECTIVE
	AD No.: 2015-0170 **Date: 18 August 2015** Note: This Airworthiness Directive (AD) is issued by EASA, acting in accordance with Regulation (EC) No 216/2008 on behalf of the European Community, its Member States and of the European third countries that participate in the activities of EASA under Article 66 of that Regulation.

This AD is issued in accordance with EU 748/2012, Part 21.A.3B. In accordance with EU 1321/2014 Annex I, Part M.A.301, the continuing airworthiness of an aircraft shall be ensured by accomplishing any applicable ADs. Consequently, no person may operate an aircraft to which an AD applies, except in accordance with the requirements of that AD, unless otherwise specified by the Agency [EU 1321/2014 Annex I, Part M.A.303] or agreed with the Authority of the State of Registry [EC 216/2008, Article 14(4) exemption].

Design Approval Holder's Name: AIRBUS	Type/Model designation(s): A318, A319, A320 and A321 aeroplanes

TCDS Number:	EASA.A.064
Foreign AD:	Not applicable
Supersedure:	None

ATA 92	**Electric and Electronic Common Installation – Cockpit Panel Bracket – Inspection**

FIGURE 16.3   An airworthiness directive issued by EASA.

```
SERVICE BULLETIN
    SUMMARY
```

This summary is for information only
and is not approved for modification of the aircraft

MANDATORY MANDATORY MANDATORY

ATA SYSTEM : 34

TITLE : NAVIGATION - AIR DATA - INSTALL SEXTANT PITOT PROBE PN C16195AA.

MODIFICATION No. : 25998P4400

REASON/DESCRIPTION/OPERATIONAL CONSEQUENCES

Operators have reported airspeed discrepancies while flying under heavy precipitations or in freezing weather conditions.

FIGURE 16.4   A service bulletin.

7/12/16   AC 21-29D

U.S. Department
of Transportation
**Federal Aviation
Administration**

Advisory
Circular

Subject: Detecting and Reporting Suspected Unapproved Parts	**Date:** July 12, 2016 **AC No:** 21-29D **Initiated By:** AIR–100

1   **PURPOSE.**

This advisory circular (AC) provides guidance to the aviation community for detecting suspected unapproved parts (SUP) and reporting them to the Federal Aviation Administration (FAA). Appendix A contains FAA Form 8120-11, *Suspected Unapproved Parts Report*, (with instructions) which serves as a standardized means of reporting. See appendix B for definitions specific to this AC.

2   **AUDIENCE.**

This AC is applicable to all personnel involved in producing, selling, and distributing aircraft parts and to all persons who perform maintenance, preventive maintenance or alterations.

FIGURE 16.5   An advisory circular.

16.2   BABEL CONFOUNDS – FORM VS. CONTENT

Literal or figurative, enough of terms and words,
A burning soul do I seek, otherwise it bores.
Kindle thy soul with the flame of desire,
Have the flame consume thought and expression, entire.

– Rumi (Translated by author)

In technical writing, form is not less important than content. How documents are laid out will determine what happens next. Finding order in chaos is not everybody's art. Jorge Luis Borges has written an amazing short story, "The Library of Babel," in which he describes a library that is infinite in extent and contains all books ever written, all human knowledge. However, it has no order. The challenge for the librarians is to find the catalog of all catalogs, the one that would include all the information found in the infinite library. That catalog can't, however, be found.

In Borges' fictional library – as in many pieces of technical documents – knowledge and wisdom that preceded the technical manuals do exist, but they're out of reach and stand for chaos and ignorance.

Although we humans are finite in reasoning and physical extension, Borges shows that we're also capable of conceiving the infinite through imagination as an idea. The lowest bidder cannot necessarily provide the highest quality. Purchasing and procurement requirements do play a significant role for the future confusions, annoyances, and frustrations. Those who are in charge of ordering or purchasing, say, computer software programs will not be using it. The procurement department is not the end user. My QA mentors – ISO9001 and AS9100 in particular – have been on the same page that if an organization is doing Section 7.0 (purchasing) properly and fully, then they're very likely to be doing everything else right within the system of processes.

16.3   OID ISN'T VOID: O-LEVEL, I-LEVEL, D-LEVEL

It is good to know that the support systems business unit of Boeing Integrated Defense Systems makes support equipment for military aircraft. The equipment is sold to all four major service branches in the United States (Army, Air Force, Navy, and Marines), as well as to foreign customers. Boeing Support Systems started in this market to support its own aircraft, but it has since expanded and tried to compete for equipment that supports other systems as well. Boeing Support Systems makes and sells both mechanical and electrical support equipment to fit one of three different levels of maintenance: Organizational Level (O-Level), Intermediate Level (I-Level), and Depot Level (D-Level).

Now let's learn what each one of these levels is about.

The Organizational Level (O-Level) signifies that the equipment tests aircraft on the flight ramp. Therefore, the equipment must be able to endure dust, rain, snow, very cold and very hot temperatures, and also humidity.

The Intermediate Level (I-Level) indicates a controlled environment with the appropriate heating, cooling, and humidity control for the tested units. This equipment is called Weapons-Replaceable Assemblies (WRAs) by the Navy and Line-Replaceable Units (LRUs) by the Air Force. To summarize, they are the repairable elements within a system that can be changed at the flight line.

The Depot Level (D-Level) includes maintenance performed at either an MRO (Maintenance, Repair and Overhaul) military facility or by the manufacturer in a controlled environment on the next lower assemblies from the I-Level. Shop-Replaceable Assemblies (SRAs) is the phrase used by the Navy, and Shop-Replaceable Units (SRUs) is used by the Air Force.

16.4   TYPES OF TECHNICAL MANUALS

My experience in working with technical publications in commercial aviation indicates that these airplanes are accompanied with 50+ manuals. For a military aircraft this could comprise of 30+ manuals:

1. Acceptance Test Procedures
2. Aircraft Characteristics/Specifications Manual
3. Aircraft Communication Manual
4. Aircraft Components Maintenance Manual
5. Aircraft Deactivation Procedures Manual
6. Aircraft Flight Handbook Manual
7. Aircraft Flight Manual
8. Aircraft Fuel Systems Manual
9. Aircraft Hydraulic Systems Manual
10. Aircraft Maintenance Manual (all procedures for maintaining and repairing a/c systems)
11. Aircraft Operating Manual
12. Aircraft Overhaul Manual
13. Aircraft Recovery Manual
14. Aircraft Refurbishment Manual
15. Aircraft Repair Manual
16. Aircraft Schematics Manual
17. Aircraft Wiring List
18. Aircraft Wiring Manual
19. APU Build-up Manual
20. Cabin Crew Operating Manual
21. Cable Fabrication Manual
22. Cargo Loading System Manual
23. Cleaning, Inspection and Repair Manual
24. Component Location Manual
25. Component Maintenance Manual (overhaul aircraft parts manual)
26. Configuration Deviation List
27. Duct Repair Manual
28. Electrical Standard Practices Manual
29. Engine Maintenance Manual
30. Engine Manual
31. Fault Isolation Manual
32. Fault Reporting Manual
33. Flight Crew Operating Manual (systems, normal operations, abnormal operations, supplementary procedures, special operations, and performance)
34. Flight Crew Training Manual
35. Fuel Pipe Repair Manual
36. Ground Equipment Manual
37. Illustrated Parts Catalog (displays subsystems disassembly and related part numbers)
38. Illustrated Tool and Equipment Manual
39. Livestock Transportation Manual
40. Maintenance Planning Document (planning for maintenance required checks and their content – Daily, Weekly, A, B, C, or D Checks)

41. Master Minimum Equipment List (explains minimum requirements of each system required to dispatch the a/c)
42. Noise Definition Manual
43. Noise Information Manual
44. Nondestructive Testing Manual
45. Qualification Test Reports
46. Quick Reference Handbook
47. Service Bulletins
48. Stress Analysis Manual
49. Structural Repair Manual (repair procedures required for a/c airframe side or structure damage)
50. System Schematic Manual
51. Tools Manual (part number and shape of aircraft tools)
52. Troubleshooting Manual (step by step procedures to detect fault reasons)
53. Weight and Balance Manual
54. Wiring Diagram Manual

16.5   ATA 100 CHAPTERS

Air Transport Association (ATA) Chapter-Section-Subject/Unit numbering system, whether in paper or digital format, is a standard numbering system used throughout most jet transport technical documentation. It follows ATA Specification Number 100 which specifies all technical data be organized by this number system. Consisting of three elements, this numbering system assigns an ATA chapter number to each aircraft system. For example, ATA Chapter 34 is for the navigation system, ATA Chapter 52 is for door systems, and so on. The second element assigns an ATA section number for each subsystem, and the third element is a unique number assigned by the aircraft manufacturer for a specific component. For example, the vane on an AOA (angle of attack) sensor, a cargo door locking system, or the fuel system temperature sensor which is used to provide a temperature indication in the flight deck (Figure 16.6).

16.6   NUMBERING COMMERCIALLY

16.6.1   System–Subsystem–Unit

Example:

ATA 34-11-00 (Navigation–Angle of Attack Sensor)
ATA 52-38-00 (Doors – cargo door) (Figures 16.7–16.10).

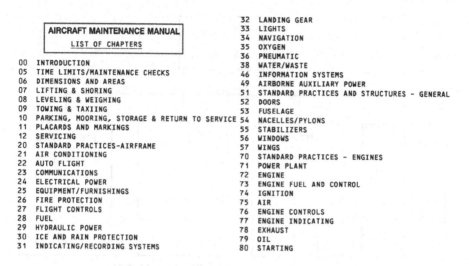

FIGURE 16.6 List of ATA chapters.

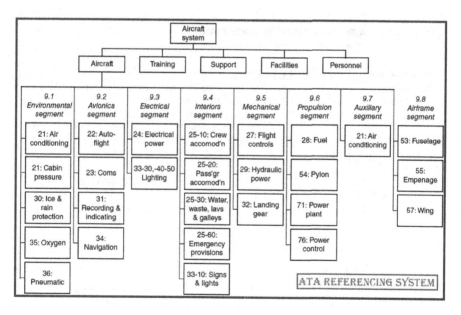

FIGURE 16.7 ATA – aircraft systems, a breakdown.

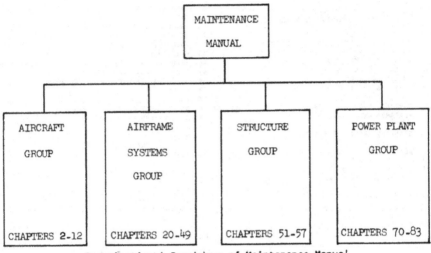

Organizational Breakdown of Maintenance Manual

FIGURE 16.8   Aircraft maintenance manuals – a breakdown.

```
AIRCRAFT MAINTENANCE MANUAL
```

2. Aircraft Maintenance Manual Numbering System

A.   The AMM numbering system divides the manual into Chapters, Sections, Subjects and Page. Each number is divided into three elements. Each element has two digits.

(1)   The Chapter and Section Numbers (Sub-Systems 10, 20, 30, etc.) are assigned by ATA Specification 100. The Sub-Sub-System number is given by the manufacturer. The third element (Subject/Unit) is given by the manufacturer.

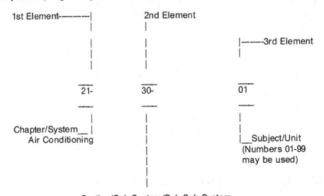

FIGURE 16.9   Aircraft manuals – numbering system.

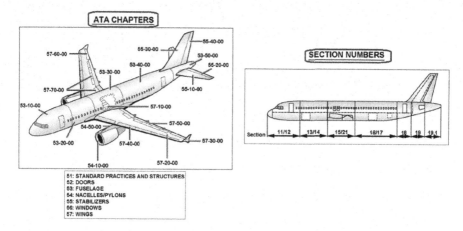

FIGURE 16.10 ATA chapters.

16.7 THE CTPL HUB

Central Technical Publication Library (CTPL) is the hub for receiving, managing, controlling, updating, and disseminating the tech pubs. The CTPL librarian manages the library and is also responsible for scheduling and conducting audits to ensure the existence of compliance and the absence of nonconformity as per the documented requirements.

17 Numbering Publications Noncommercially

According to NAVAL AIR SYSTEMS COMMAND TECHNICAL PUBLICATIONS LIBRARY MANAGEMENT PROGRAM (NAVAIR 00-25-100) (www.navybmr. com/study%20material/NAVAIR%2000-25-100%202016.pdf), technical manuals (TMs) are divided into two major types: operational and maintenance. Operational manuals are manuals and other forms of documentation that contain a description of a system as well as instructions for its effective use. Maintenance manuals are documents containing a description and instructions for the effective use and support of the very system from a viewpoint of repair.

In the noncommercial sector, there's two types of tech pubs: general – for all aircraft types – and type model series – for a specific type.

According to http://navyaviation.tpub.com, there are two numbering systems presently in use by Naval Air Systems (NAVAIR): the older NAVAIR publication numbering system and the newer Technical Manual Identification Numbering System (TMINS) which, according to the Aviation Maintenance Administrationman Basic (http://navybmr.com/study%20material/NAVEDTRA%2014292.pdf) is part of the effort to standardize technical manual numbers for all ships, aircraft, and equipment.

You must be able to use both numbering systems. The TMINS assigns each technical manual a unique identifying alphanumeric designation patterned after the 13-digit National Stock Number (NSN); for example, A1-F18AA-NFM-500. It serves as the technical manual identification number. Additionally, TMINS contains a provision for adding a suffix to give the security classification and other information considered important.

According to Aviation Maintenance Ratings and Standard Technical Manual Identification Numbering System (TMINS) – released on May 14, 1980, and available in the public domain – published by the US Navy (http://navybmr.com/study% 20material/navedtra%2014022.pdf), the NAVAIR manual publication number consists of a prefix (NAVAIR or NA for NAVAIRSYSCOM) that designates the command responsible for developing or maintaining the manual. The manual number consists of three parts, separated by dashes (-). Additional numbers may be added to show multiple volumes of a manual.

As per TMINS, the following figures present an example of the TMINS numbers assigned to a family of aircraft-related TMs for large, intermediate, and small aircraft (Figures 17.1–17.3).

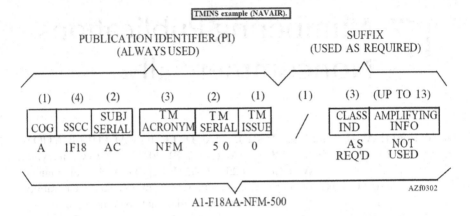

FIGURE 17.1 Numbering in TMINS.

Organizational Maintenance Series:

Work Unit Code	
A1-F18AA-110-XXX	Airframe Maintenance
A1-F18AA-130-XXX	Landing Gear System
A1-F18AA-270-XXX	Turbo Fan Power Plant and Related Systems
A1-F18AA-460-XXX	Fuel System
A1-F18AA-540-XXX	Telemetry System
Volume Breakout	
A1-F18AA-540-100	Principles of Operation
A1-F18AA-540-200	Testing/Troubleshooting
A1-F18AA-540-300	System Maintenance
A1-F18AA-540-400	System IPB
A1-F18AA-540-450	Master System IPB Index
Special Breakout	
A1-F18AA-540-500	System Schematics

Maintenance Requirement Series:

A1-F18AA-MRC-000	Periodic Maintenance Information Cards - General
A1-F18AA-MRC-100	Aircraft Turnaround Checklist
A1-F18AA-MRC-200	Daily Servicing/Special Cards
A1-F18AA-MRC-300	Phased Package Sequence Cards
A1-F18AA-MRC-400	(others as required)
A1-F18AA-WUC-800	Work Unit Code Manual

Structural Repair Manual Series:

A1-F18AA-SRM-000	Structural - General - Unclassified
A1-F18AA-SRM-100/( )	Supplement - Classified
A1-F18AA-SRM-200	Corrosion Control
A1-F18AA-SRM-300	Non-Destructive Inspection
A1-F18AA-SRM-400	Illustrated Parts Breakdown (IPB)
A1-F18AA-SRM-450	IPB Master Index
A1-F18AA-IPB-450	Master Aircraft IPB Index

FIGURE 17.2 TMINS, a breakdown.

Aviation Maintenance Ratings NAVEDTRA 14022 and Aviation Maintenance Administrationman Basic NAVEDTRA 14292, both available in the public domain, talk about the tech pubs layout (structure) as well as the numbering system.

According to NA-00-25-100 (NAVAIR Technical Publications Library Management Program), there's a conventional numbering system that consists of a prefix and a combination of numbers and letters divided into three parts and separated by dashes (NA-01-…) and a Technical Manual Identification and Numbering System (TMINS) developed in coordination with other systems commands. TMINS may be accessed on the world wide web.

The structure of numerical and alphabetical combinations of a NAVAIR technical manual number identifies the basic equipment category, main groups within the category, specific item of equipment, type of usage, type or model designation, and specific type of manual.

Part I of the publication number is the category. Normally it is a two-digit number (in some cases two digits and a letter). It designates the major category of the manual; for example, 00 tells you that this is a general manual; 01 is for airframes;

TMINS	MANUAL DESCRIPTION

LARGE AIRCRAFT

Maintenance Requirement Series:

A1-F18AA-MRC-000	Periodic Maintenance Information Cards - General
A1-F18AA-MRC-100	Aircraft Turnaround Checklist
A1-F18AA-MRC-200	Daily Servicing/Special Cards
A1-F18AA-MRC-300	Phased Package Sequence Cards
A1-F18AA-MRC-400	(others as required)
A1-F18AA-WUC-800	Work Unit Code Manual

Organizational Maintenance Series:

Work Unit
   Code ⌐

A1-F18AA-110-XXX	Airframe Maintenance
A1-F18AA-130-XXX	Landing Gear System
A1-F18AA-270-XXX	Turbo Fan Power Plant and Related Systems
A1-F18AA-460-XXX	Fuel System
A1-F18AA-540-XXX	Telemetry System

Volume Breakout

A1-F18AA-540-100	Principles of Operation
A1-F18AA-540-200	Testing/Troubleshooting
A1-F18AA-540-300	System Maintenance
A1-F18AA-540-400	System IPB
A1-F18AA-540-450	Master System IPB Index

Special Breakout

A1-F18AA-540-500	System Schematics

FIGURE 17.3   TMINS – maintenance and organizational.

and 02 is for power plants. (For a complete breakdown of publication numbering categories, refer to Naval Air Systems Command Technical Manual Program, NAVAIR 00-25-100.)

Part II of the publication number is made up of numbers (or numbers and letters). They identify either a basic aircraft model, the manufacturer, or the specific class, group, or subcategory of the manual. For example, in the following figure, the number F14AAA in view A identifies the aircraft model. In view D, 75PAC identifies Lockheed as the manufacturer of the P-3C airframe (Figure 17.4).

Part III of the publication number usually identifies a particular type of manual. For example, -1 identifies the NATOPS flight manual, -2 the maintenance instruction manual, -3 the structural repair manual, and -4 the illustrated parts breakdown.

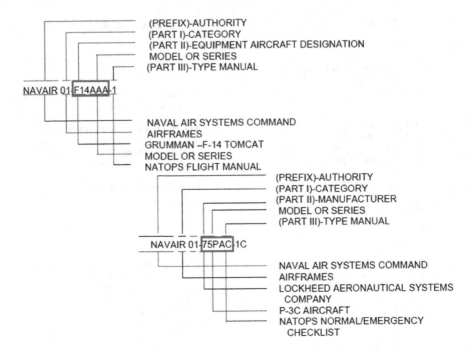

FIGURE 17.4 Aircraft model and manufacturer.

Additional numbers may be used to show system grouping breakout by volume or subsystem grouping by sub-volume. For example, in the number -2-2, the second -2 indicates the second volume of a maintenance manual. In the number -2-2.1, the .1 indicates a sub-volume within the grouping.

17.1 E FOR ERA: INTERACTIVE TECH PUBS

Paper shuffling and paper chasing era has been losing color rapidly and is being replaced by the software application or other process, digitizing paper technical manuals, or upgrading to electronic technical manual in an interactive manner. I am sure that like me, you have also seen CDs that contain tech pubs in electronic formats with good search capabilities to locate the specific info you're looking for. Interactive electronic technical manual (IETM) is being widely used these days. In the civilian sector, Airbus World is a good example of a platform created by Airbus that can be accessed through legitimate login credentials.

17.2 DISCREPANCY, SEE?

TPDR (Technical Publication Deficiency Report) can be used to report technical publication deficiencies. Updates are issued through the Rapid Action Changes/Interim Rapid Action Changes (RACs/IRACs), Formal Changes, Revisions, or NAVAIR TM Notices. Technical manual deficiencies, typos, or errors can be reported via

ABC Company

TECH PUBS CUSTOMER SURVEY FORM

To help us to improve the quality of our Technical Publications,
ABC invites its customers to answer this customer survey form.

| For ABC use only |
| Date received : ___/___/___ |
| Ref No : _____ |

Publication identification

ATA Nr : 34-20-12	Revision : 18	
Title : ANGLE OF ATTACK SENSOR		Aircraft : A300-600, A318, A320, A340

Customer Information

You are : ☐ **Engineer** ☐ **Librarian** ☐ **Instructor** ☐ **Other** (please specify) : ....

Name (optional) :	Company Name :
Address :	
Phone No :	E-Mail address :

Customer Satisfaction

Please tick the appropriate boxes corresponding to your opinion about this publication :	☺☺ Very Good	☺ Good	☹ Needs some improvement	☹☹ Needs much improvement
Publication accuracy	☐	☐	☐	☐
Completeness of information	☐	☐	☐	☐
Ease of use of illustrations or tables	☐	☐	☐	☐
General appreciation	☐	☐	☐	☐

	Yes	No
Do you use this publication for training purposes?	☐	☐

	☺☺	☺	☹	☹☹
How do you consider this publication with regard to other publications you use?	☐	☐	☐	☐

Please let us know your remarks, comments and suggestions about this publication
_____
_____
_____
_____

Please send us this form by fax or email:
E-Mail: techpubs@abc.com   Fax: 1-800-123-4321

FIGURE 17.5   Tech manual user's feedback form.

Form 4790/66 (Technical Publication Deficiency Report) to the Naval Air Technical Services Facility (NATSF). Also, the NAMDRP (Naval Aviation Maintenance Discrepancy Reporting Program), managed by the QA division, is a method for reporting such deficiencies. The QA initiates and maintains a report control number (RCN) to each NAMDRP report (Figure 17.5).

17.3   PUBS FOR THE MOBS: THE ART OF SUPPLEMENTING

There are also smaller versions of tech pubs available which are more convenient for users than one large manual. These documents – functioning as supplements – can be produced one at a time, as information becomes available. Being also easier to update, these documents can be added without having to reorganize, reprint, or distribute large manuals.

If created and used properly, supplemental materials can enrich a user's engagement and understanding. Such materials allow users (readers) to focus on specific parts or areas of a longer process or document. Being also local technical directives that enhance clarity, they supplement technical publications when there's substantial change in a procedure, an instruction, or a requirement. In many cases, these supplements provide information, instructions, material substitutions, and test or repair procedures that are not otherwise available.

When properly documented and compliantly implemented, safety of the liveware, hardware, and environment will be maintained. For more information regarding such directives, you may refer to the available resources (e.g. NAVAIR 00-25-100).

18 Writing in Progress
Instruments at Work

People come and go. They transfer, promote, retire, or get laid off. But documents stay. Requirements survive. They evolve. They do not retire with people who created them or used them.

In his famous play, *Six Characters in Search of an Author*, Luigi Pirandello narrates the (fictional) story of six characters who wander into a theater rehearsal, explaining that they have been abandoned by their author:

> FATHER [moving her aside]. Yes, lost, that's the word for it! [To the DIRECTOR without a break] Lost, you see, in that the author who created us alive either wouldn't or in practice couldn't bring us into the world of art. And that truly was a crime, sir, because someone who has the luck to be born as a living character can laugh even at death. He never dies. The man will die, the writer, the instrument of creation; but the creature never dies!
>
> **(Pirandello 2019)**

As Pirandello defines the writer on the above passage in his play, s/he is only an "instrument," a function, a facilitator. Your reader's life does not depend on you as a writer, but on your creation as a phenomenon that could outlive you. Create creatively.

18.1 SPEAKING FIRST, WRITING NEXT?

18.1.1 LOGOCENTRISM DECONSTRUCTED

In his 1873 essay "Über Wahrheit und Lüge im außermoralischen Sinne" ("On Truth and Lies in a Nonmoral Sense"), Friedrich Nietzsche (1844–1900) argues that we necessarily and often unknowingly use metaphors when we discuss the question of truth, taking them to be the original things themselves:

> When we talk about trees, colors, snow, and flowers, we believe we know something about the things themselves, and yet we only possess metaphors of the things, and these metaphors do not in the least correspond to the original essentials.
>
> **(Nietzsche 2010)**

In one of his most influential early books, *Of Grammatology*, Jacques Derrida, the French philosopher who wrote extensively on language and meaning, complains that Western intellectual culture and history have been obsessed with speech rather than writing. On this account, truth emerges from face to face dialogue, whereas writing is one step removed from the original act of speaking. There is a binary opposition between speech and writing, and, according to Derrida, writing has been the repressed side of this opposition ever since Plato. The privileging of speech over writing, with the idea of the speaker being present in person, Derrida equates with "logocentrism" (logos is the Greek term for "speech, word, law, or reason"). Moreover, the priority of speech over writing creates the illusion that words (signs) refer to concrete objects that are directly present to speakers who enjoy privileged access to them.

Derrida believed that we put far too much emphasis on the significance speech – he calls that emphasis *phonocentrism*. He also coined the term *logocentrism* to demonstrate the belief that signifiers (words, images) can contain the essence of their signifieds (logos is ancient Greek for: word).

18.2   EASY ENGLISH WOULD BE EASILY HARD

You lovers of the English language would tremendously enjoy the mysterious two-letter word that perhaps has more meanings than any other two-letter word, and that is "UP."

Metaphor is at the very heart of language (and the word "heart" is itself a metaphor here – as is language, which comes from the Latin for "tongue").

The bandage was wound around the wound.
The farm was used to produce produce.
The dump was so full that it had to refuse more refuse.
He must polish the Polish furniture.
He could lead if he would get the lead out.
The soldier decided to desert his dessert in the desert.
Since there is no time like the present, he thought it was time to present the
    present.
A bass was painted on the head of the bass drum.
When shot at, the dove dove into the bushes.
They did not object to the object.
The insurance was invalid for the invalid.
There was a row among the oarsmen about how to row.
They were too close to the door to close it.
The buck does funny things when the does are present.
A seamstress and a sewer fell down into a sewer line.
To help with planting, the farmer taught his sow to sow.
The wind was too strong to wind the sail.
Upon seeing the tear in the painting, I shed a tear.
She had to subject the subject to a series of tests.
How can I intimate this to my most intimate friend?

Let's face it – English is a crazy language. There is no egg in eggplant, no ham in hamburger, no apple and no pine in pineapple. English muffins weren't invented in

England or French fries in France. Sweetmeats are candies while sweetbreads, which aren't sweet, are meat.

And why is it that writers write but fingers don't fing, grocers don't groce, and hammers don't ham? If the plural of tooth is teeth, why isn't the plural of booth, beeth? One goose, two geese. So one moose, two meese? One index, two indices? Doesn't it seem crazy that you can make amends but not one amend?

If teachers taught, why didn't preachers praught? If a vegetarian eats vegetables, what does a humanitarian eat?

18.3   A STILL MIND CAN STILL WRITE

When happening to deal with a child, the childlike language need I mind.

– Rumi (translated by author)

Here are some tips that I have learned and practiced throughout my career as a non-native English speaker writing in English:

- Choose your words with precision and care. Every single word counts.
- Keep a thesaurus handy. Think through all the possible inflections of meaning a word offers.
- Verbs convey action – do not underestimate their role and power.
- Watch out for the passive voice. Use the imperative mood and the active voice. In an active voice, the subject acts upon the verb, but in a passive voice, the subject receives the action (of the verb). "To be" is an infinitive with variations. Always ask yourself who or what is the source of the action (verb).
- Avoid run-on sentences or merging long sentences together without proper punctuation. Run-on sentences make the rhythm disappear.
- Your reader's attention is a gift in this busy life and you shouldn't take it for granted. Attention is hard to gain, but very easy to lose.
- Read your writing aloud to yourself or to another person. You'll catch the rhythm.
- The target audience – have someone read through and interpret your work. Test your instructions by having someone else follow them while you observe them.
- Brevity – use short sentences and simple present tense, as frequently as possible.
- Write for people. From a human being to human beings – tell them what, show them why.
- Jargons – avoid jargons unfamiliar to your readers. If you're using abbreviations and acronyms for the first time, spell them out.
- Elegant variation – use two different words for the same concept. Variation helps with understanding.
- Eliminate ambiguity. Be as clear as possible. A clear mind precedes a clear writing.

- Illustrations – use them as required. A picture can speak a thousand words.
- Do not burden your reader with the tedious task of hunting for antecedents to your pronouns.
- Let your text cool off and then attend to it later for fine-tuning and pruning away of things that simply do not add up. Cut out sections that deviate from your central intent.
- Care for your words. Words count. They are going to be speaking on your behalf to your reader. Be the writer you wish to be a reader for.

Rumi tells us stories – layered narratives – in his *The Masnavi*. One of the stories he tells us is the story of a merchant who had a parrot in a cage. Ready for his trip to India, the merchant asked the parrot if she had any messages for her kinsmen in India. The parrot asked him to tell them that this is not fair at all for her to be confined to a cage while they are enjoying their freedom. The parrot asked the merchant to tell them to raise their glasses for her in the rose garden they live in. The merchant didn't think much about the parrot's demand, but promised to do as the parrot pleased.

Upon hearing the parrot's message of grief, one of the parrots began to shiver, fell off the branch of the tree he had perched on, and dropped dead. The heartbroken merchant became very sad to have been a messenger of sadness and death. When the merchant returned, he shared the sad story with his parrot. As soon as the parrot heard the tale, she fell down dead in her cage. The grief-stricken merchant took out the corpse of his beloved and threw it away. To his shock, the dead parrot flew away. When the astonished merchant asked her what had happened, the parrot told him that the parrot in India had communicated a secret message to her that held the key to her freedom. Then the parrot bid her merchant master farewell for the last time and quickly flew out of sight.

Moral: learn the secret language of your loved ones if you do not wish to lose them to a fake death.

My fellow writers, in order not to lose your readers, spend some time to tap into their minds and learn how they communicate secretly. They might be communicating against you if you fail to learn their language.

18.4   INFINITY CONFUSES INDEFINITELY

In his 1975 short story, "The Book of Sand," the Argentine writer Jorge Luis Borges narrates a story in which a stranger who happens to be a Bible seller shows up at his door. Offering him a clothbound book from his belongings, the stranger mentions that it is an infinite book which is mysteriously designed and is different every time you open it. Like sand, it has neither a beginning nor an end. The narrator purchases the book and his obsession with it grows alarmingly. Refusing to leave his apartment and failing to master it, he becomes a prisoner of the book. Soon he realizes that the book, with its random pagination, initially brought joy, but turned into a nightmarish, horror being that was destroying his life as well as reality. Thinking of burning it, he ultimately decides to take it to the library and buries it in a dusty corner like a needle in a haystack.

Besides the other beauties that a reader may find and appreciate in this piece, there is an important message to both writers and readers in this parable by Mr. Borges. You may be granted access to infinite knowledge and eternally growing data, but with no clear instructions, structure, formatting, and guidelines being provided on how to use it, you'd end up giving up and think of possible ways to get rid of that vast, infinite amount of knowledge.

18.5   EXPERIENCE COUNTS

18.5.1   RECOMMENDATIONS FOR ME, FOR YOU, FOR US

Blessed are those souls who have had mentors who show them where to look at, but won't tell them what to see. Directions matter. No words are to express my gratitude to my generous, welcoming mentors, colleagues, fellows, and students who inspire me every day. Among them is John Doherty, a retired B747 Captain. During his career he accumulated over 22,000 hours of military and airline flying in domestic and international operations. He served as a director of B747 Flight Training at a major US airline. In that position he was instrumental in developing leading edge pilot technical publications and Advanced Qualification Programs with a pioneering emphasis on human factors. Post retirement he consulted on pilot procedural documents and pilot training with a number of airlines and medical transportation companies.

According to Captain Doherty, when an organization contemplates either writing new technical publications or upgrading existing publications, a good starting point is to determine which of the following would be a guiding principle:

1. The publications will be of high quality that are used by and valued by the line workers who are affected.
2. The publications are intended to meet regulatory, insurance, liability, or other concerns; usability by line workers is not a primary concern.
3. The publications are developed as inexpensively as possible to meet minimum requirements.

The guidance here is applicable to only number one above. This particular guidance is specific to pilots and the airline environment, but the concepts can be applied to other endeavors.

Quality publications can only be developed if there is strong support from the top of the organization. Top leadership will understand that quality costs more in the short run but pays dividends in terms of safety, avoided liability, quality of product, profitability, customer satisfaction, and employee retention.

A statement of work should be created setting out vision, scope, costs, reporting lines, resources, budget, and makeup of the team responsible for doing the actual writing. The urge to jump in and start writing should be resisted until these things are understood and approved by leadership.

The writing team needs to be led by a Subject Matter Expert (SME) who has a deep understanding of the aircraft systems, how to operate the aircraft in the "real world," and has an understanding of the pilot demographics, how actual pilots in that environment think and work, both individually and as teams. It's a big help also if the SME has a degree of technical writing skill as well as other communication skills. The SME should have a talent for making the publications user-friendly for pilots in the real world.

There needs to be a technical writer who:

- Works *side-by-side* with the SME. A technical writer with enough aviation background to be able to understand the nuances of what he or she hears from the SME.
- Can write in simple, intuitively organized, straightforward language that can be easily absorbed, followed, and valued by the pilots.

A part of the team should also include a publishing technician, a person who has a deep understanding of using the software that manages the actual document. This is the person responsible for getting the drafts from the technical writer into the organization's pubs system, setting publication standards, conventions as to things like acronyms, font size, page numbering, etc., and in getting everything into print in a form that is familiar across the organization. Consistency of conventions throughout the publications is paramount. No particular need for this person to be conversant in flight operations procedures.

All proposed procedures should be extensively field tested, at first in simulators, then with the FAA approval in actual line flying. This line flying should be accomplished by line pilots with extensive experience and specialized training and approvals for using these tentative procedures in actual line flying. Requires close coordination with the FAA.

Also critically important to a successful publication: the introduction to the pilots and the training on "how to use" for the pilots. Typically, a major change in publications is known months ahead of time, and pilots should be prepared over time for that change, the reasons, an understanding that there is line pilot input, etc. If there is a union involved, the union should be brought into the conversation early on and throughout.

Once completed, pilots and other end users should be trained in the use of the new manuals. And it is inevitable that there will be some shortcomings or errors when the new publication is put into use. A system for soliciting and tracking those items should be in place ahead of time, and it should be used in early updates and corrections. In particular, end users will notice that their concerns and suggested corrections are being paid attention to and that will increase their confidence in the publication.

If pilots don't have confidence in the publications, or if they find them awkward to use, they will start developing their own ad hoc practices, and this is of course detrimental to overall safety.

Let's review the words of wisdom from some distinct scholars concerning the written form of communication.

Dr. Earl A.

Technical writing ... and why is most technical writing so bad? I think it might be that the companies that need good technical writers aren't willing to spend the money to hire them, that's all. Most "tech" people are lousy writers and the main problem is that they lack explicitness. Also, they seem to be unable to put themselves in the place of a reader who needs step-by-step instructions. In general, a good writer is also an avid reader.

John D.

Brief background: in the late 1980s, XX Airlines merged with YY requiring merging of flight operations policies and procedures. Two different cultures. The managing director of flight operations at that time made the inspired decision to redo our flight operations manuals from scratch rather than try to cobble the two carriers' procedures together. I was responsible for creating the ZZ manuals, but we also undertook to standardize all the manuals across seven different aircraft types. I say this in all humility, the work that my team did on the ZZ had a profound impact on all the other aircraft manuals via the standardization process. The project took three years across the airline, and in my once again humble opinion was tremendously successful.

You can easily imagine that I can and have talked for weeks about the process. I will try to summarize in a few brief points what I think are the critical factors. Whether a corporation will pay for an effective writing team depends on the corporation. A lot of tech writing in the United States is CYA. A corporation's legal department wants to be able to say "we told you" in case something goes wrong. That is also the cheapest kind of tech writing. Quality costs more originally (many economists say about 25% more) but it pays off over time in terms of quality of the product, reduced legal liability, employee retention, customer satisfaction and loyalty, etc. Personally, I wouldn't work for a corporation that is focused on liability over quality. It's a false dichotomy in every way except for the short term. Focusing on liability over quality more or less ensures that money not spent on quality will later be spent on defending against liability.

Dr. Karen R.

While I do not know much about "technical writing," as somebody who has been teaching college composition classes for over 15 years, I have seen a decline in college students' abilities to write well. I often ask myself questions similar to the ones you do. Why are my native English speakers having such a difficult time writing in English?

I do not think there is a simple answer to that question, although I wish there was!

I think the main issue stems from the fact that most students do not speak standard English. Students use a lot of slang, and they make a lot of grammar errors in their speech (double negatives, subject-verb agreement, misused pronouns, lack of parallel structure, misplaced modifiers, etc.). When students write like they talk, their prose (if I dare to call it that – ha!) contains the same errors as their speech.

I think social media platforms have also negatively affected students' abilities to write well. These platforms encourage people to condense sentences and abbreviate words, and I have started to see some of that in my students' writing.

I encourage my students to READ more because I believe that's the best way to learn about how language works. I also encourage students to take a grammar class. I think all colleges offer grammar and basic composition classes.

Dr. Tim M.

Writing is hard and most people don't want to do it in any language. Hence, the dream of voice recognition, voice-to-text interfaces, digital solutions bypass the written word. Even the technical-communications faculty that I work with would rather produce videos and send voice messages than compose written prose texts.

Universities are under pressure to produce graduates as a quality metric. So they've made it easier and easier to avoid writing. Lots of college majors involve almost no writing at all, and there is a "lot" of cutting and pasting and outright plagiarism. Being a native speaker doesn't guarantee good writing if you never practice writing.

My reaction when somebody wants to send me an instructional video is to say "Can I just have a document?" but I think for every person like me (and I guess you too) there are a dozen others who hate the thought of reading and would rather have somebody or some video talk them through a process. As to the native-speaker issue … I think that native speakers of most languages rarely come under pressure to refine their communication in that language; but that pressure drops to zero for native speakers of English. We expect the whole world to use English with us – the cliché of the American traveling abroad whose opinion of different places is based on how much English they can use there. … It is a cliché but it exists, and educated people perpetuate it.

Dr. Paul L.

It goes without saying that any tech pub needs to be as clear as possible. But it also has to induct the learner into a professional discourse (the jargon of the job, if you like). This jargon might be off-putting to the outsider or the beginner, but it's obviously vital to those doing the job, and indeed (as you say) a matter of life and death. Most professions have their jargon: it's a sort of shorthand for those that need it, and it's got to be learned and used as a matter of professional normality. You drop it when you go home, but it's a vital shared conduit at work. That's my only reflection, really. But it's meant to indicate that jargon (which may not be everyday language) has its uses! It's the language of the professional tribe.

Writing at degree level, in my experience teaching it (for 20+ years, 1990–2014) in an English university, is quite unsatisfactory. The grading is often not exacting enough (so students tend to think of it as a soft option) and the extent to which it overlaps with any literature degree is not sufficiently acknowledged, i.e. in order to write well you need to read widely and well, though in a different way and to different ends from those of the literary scholar. At least this is so if you are studying creative writing at university. If all you want to do is write (usually novels), then it's pointless going to university. Save yourself the money! As it is, most students only pretend they believe you can learn to write. One can tell this from their work and reflection on it. And in practice they are right: mostly they haven't learned anything. This is partly teachers' faults. Teachers have their own learning to do: reading needs to be "angled" in a creative way. Students will only learn from a text if it is taught in a way that is different from the normal literary-critical way – especially nowadays when literary study is emphatically "historical," contextual, and theme based.

If you just arrange for writing students to attend literature classes that they are interested in, they won't learn anything much, except on their own initiative. If they are far along in their study, they need to formulate as clear an idea as possible of what technical (that is, formal and stylistic) problems they might encounter. If they don't do this, they will just be reinventing the wheel the whole time. Most technical problems have already been solved (the tutor should be knowledgeable enough to know where and by whom), so they should be pointed to helpful texts. This can happen as their work goes on, changes, and morphs into something different. They should be pointed toward bad writing as well as good and genre-writing as well as literary writing. They need to be educated in possibilities. This is mainly prior to writing. It's only during the process and after drafts are finished that the tutor can then help with comments, that is, *teach*.

Weed

I'm not sure why technical information is often so bad, but early computer manuals were notorious for it. Certainly if people don't have a clear understanding themselves, then it will be hard for them to write clearly for others. Presumably this is mainly because most people only see things from their own point of view – if it's clear to them, then it should be clear to everyone.

When products are being developed for market, everything is working in parallel – the hardware, the software, and the documentation – so sometimes the documentation is being written while the product is still in development, and what is written won't always apply to the final product. And of course there are deadlines to meet. There is usually not enough time to give the instructions to a group of people and see where any confusion arises, then go away and rewrite, return, and retest with another group of people, and continue doing this until all the problems are ironed out. This of course is the only way to get crystal clear instructions.

One day at school, our English master got up from his desk and went to the blackboard (teachers used white chalk in those days!) and drew a simple pattern. He then said that he was going to rub out the pattern, sit down again, and the class had to tell him exactly what to do to repeat his actions. It took the rest of the lesson for us to give him the instructions with enough precision so that he redid exactly what he'd done the first time in less than ten seconds!

When I was helping to write simple text adventure computer games in the early 1980s, I would get friends who didn't know anything about the game to sit down and play it, and I'd write down anything that confused them or that they didn't understand. Then I'd go back to the program and change the parts where there had been problems for the player. After doing this at least ten times and making many changes people started enjoying the game much more. Later a reviewer in one of the computer magazines commented on one of our games how much better it flowed than most text adventures of the time, where much of the difficulty came from trying to discover how the game worked and what commands were appropriate.

I think there is (or was) a rule of thumb that programmers should devote at least a third of the time allocated on a project for testing. One job that I had was to test a new system for an electronics manufacturing company which was beginning to computerize. I was presented with the first module which was said to be fully working and ready to be implemented. This seemed doubtful because the programmers, though academically qualified, were quite young and inexperienced in writing software for use in a work environment. It took me just a few minutes to find a simple way of crashing it, which goes to show how easy it is for us to assume that everyone sees things the same way!

Dr. Margaret M.

I also find it amusing that I have all these friends from India who can speak and write fluently in English, Hinglish, Hindi, and Tamil. And they speak, but can't write Punjabi (often Punjabi for some reason), even though it only takes a few hours to learn an alphabet. And I couldn't agree more that unclear thinking is behind unclear writing. Different parts of the brain are used in speaking and writing, so there are instances of people with brain injuries who can't talk but who can write and vice versa. Strange, no?

Karen M.

Writing is a skill that must be taught. Also how you write depends on what you write and for whom. It's not easy for anyone. Only the motivated are willing to put the time in to learn. Unfortunately, many teachers are themselves deficient.

Dr. Cheryl G.

Students rarely value learning how to write until it's too late!

J. Blaylock

Your question is interesting. When I was first hired to teach at XX Community College back in 1976, I had a nice chat with my department chair. When I was walking out, he called me back and said that I was obviously excited about language and teaching. He told me to remember that 95% of my students would come into my class with a 12-year hatred of English classes. I didn't understand that, because English classes were, with few exceptions, the classes that I particularly enjoyed. His comment turned out to be largely true, and I wondered why – what was wrong with an educational system that failed to motivate learning to that degree. My goal was to change things for the better, which had a lot to do with the way I decided to teach and also with my belief that the system was often blind and that I had to circumvent the system when I thought it necessary.

That doesn't mean that I was successful, but it did mean that I've remained a happy teacher, and that I don't feel a lot of negativity from my students in regard to the subject matter of my classes. So, one small thought, is that students do not learn what they cannot attend to. They pay no attention through 12 years of English classes, thereby failing to learn to use the language more competently, and then arrive in college thinking that they can make up for those lost years by taking classes in technical writing or business writing or even literature. But they cannot. They simply do not have basic skills.

A couple of interesting things: I've found in my 45 years of teaching, that students who write good academic papers often cannot write fiction. They attempt to create a "style" and ruin their sentences. Over the years, solidly good teachers and editors I've known have attempted to become fiction writers. They're very good at telling others how to write, but somehow they cannot succeed. A talented teacher is not necessarily a talented writer. A talented poet cannot necessarily write good stories. A talented fiction writer might be an awful poet. Why?

As for tech writing, I can't stand to read it. It seems to me that the writer often makes assumptions about the reader's skills that are way out of line. Take me, for example. I have no understanding or memory for computer jargon. So two days ago when I read "open your document in Acrobat and follow the directions," I had no idea what it meant. Acrobat? I looked in my computer's list of applications and Acrobat was not listed. I'm not even sure what "open your document" means. If I open a document, it appears in Word, not Acrobat (or whatever). A tech-savvy person would assume that I'm an idiot and perhaps is correct. But his or her instructions are meaningless to me. Jargon is a curse.

One thing that's true is that American idiom is full of impossible to understand phrases, at least for nonnative speakers. I'm astonished, however, that many native students in university writing classes have no knowledge of American idioms that are more than a few years old. This problem arises from two things, I think: first, people think they want to write, but they do not listen or read or retain language. In fact, they have no interest in literature written 15 or 50 or 500 years ago, because they believe it's "out of date." Anything "old" is useless to them. You came from a literature background, with a high regard for the literature of at least two different cultures. You've paid attention to what you've read, and (as I remember very well) you have an innate enthusiasm for word play, turns of phrase, subtextual meaning, etc. But that's not as common as it should be.

I'm going to guess that people who teach tech writing often know all about the technology but nothing about writing. (And many writers – me, for example – know a lot about writing but nothing about technology.)

I'll finish by saying that I no more "know how to teach" today than I did 45 years ago.

Tim B.

I think you're living proof that being a native speaker has little to no impact on your ability to write well. I assume it transcends language. But your point gets to the root of something that isn't taught in school today, and that's decent writing. There's plenty of opportunity to let students parrot back a subject they've read about, however. In college, I studied the ways instructors wanted things delivered more than the actual writing process.

"Proper writing" is taught, sure. Make sure you follow the rules, get your tenses and sentence structure correct, engage in lively debate on the use of the Oxford comma. But none of that teaches creativity or art.

I think good technical publications are a challenge because you not only have to be both concise and accurate, but also ensure that your words are understandable by your entire audience. To me those are two completely different tasks requiring independent skill sets. Your own personal bias and experience on the subject have to be removed as well.

I write manual documents and it's difficult. But I do one thing at the end that helps quite a bit. We have a guy on staff who is a retired ... and I don't release a thing until he reads it.

He's not a particularly good writer, nor is he an engineer. But he's "one of them," closest in experience and capability to our target user. I can't tell you how many times I've spent hours structuring the steps of a difficult process, laboring over every word to make it as simple and meaningful as possible, and he's picked it up and shown me three things I've assumed that would immediately confuse or even alienate a user.

The bottom line in my humble opinion is that because tech pubs require successful application of technical skill, mastery of language, and knowledge of the target audience, they're almost doomed from the start. And they're often treated as an afterthought, not resourced anywhere close to what's actually needed.

Dr. Jerry B.

Some part of me thinks that good writing is rather more the product of one's genes than it is of one's schooling, but I've often thought that people who write well are also voracious readers –basically, they live with language. I taught at the U. of XX for 15 years and had many gifted students who were in the famous XX Writers' Workshop, where writing is not so much taught as practiced in the forms of poetry, fiction, drama.

Then there's the problem of "technical" culture or communities that have their own "idiom" or "argot" or "forms" of language. Perhaps one should think of writing in relation to communities – poetic communities, newspaper writing, writing in the business world, and of course now the "digital universe." This last seems to me one worth some serious thought – what becomes of language and the use of language in the digital world, which is rather more visual than verbal?

Insufficient distinction has been made between teaching and showing, perhaps because of their identity in ostensive definition. I can teach you how to play chess even if we have neither board nor pieces, but I cannot show you without them. I could explain to you how a certain passage of piano music should be played and then show you on the keyboard. Telling you how to use a word is not the same thing as showing you how to use it, just as describing its correct use is not the same as using it correctly. In each case I may be able to do the latter without being able to do the former.

(Bryan Magee, *Facing Death* 1977)

To say more than human things with human voice,
That cannot be; to say human things with more
Than human voice, that, also, cannot be;
To speak humanly from the height or from the depth
Of human things, that is acutest speech.

(Wallace Stevens, *The Collected Poems* 1971)

It is cold in the scriptorium, my thumb aches. I leave this manuscript, I do not know for whom; I no longer know what it is about: stat rosa pristina nomine, nomina nuda tenemus.

(Umberto Eco, *The Name of the Rose* 1986)

Appendix – Tech Writing, Sample 1

Aircraft Operations Safety
Airbus A320 Lower Lobes
Cargo Operations

INTRODUCTION

Settling into your seat for flight you will notice machinery and ground staff loading the aircraft with baggage and cargo. Most cargo operation areas are accessed from the right-hand side of the aircraft where containerized or bulk loads can be un/loaded. Bulk loads are defined as loose loads, baggage, or freight which can be loaded within a net section. Besides luggage, other loads can be found in the belly of an aircraft; therefore, passengers may be sitting over live animals, coffins, HAZMAT, or tons of green onions. Cargo holds are pressurized all the time, but are unheated in some aircraft – that's why live animals cannot be shipped on such aircraft during the winter. Pilots are always notified about Special Loads (live animals, perishables, human remains, HAZMAT, etc.) onboard through the NOTOC (notification to captain).

LOWER LOBES – AIRBUS A320

There are two lower holds (FWD and AFT) in most narrow-body aircraft, the Airbus A320 included. These holds are divided into cargo compartments. The FWD hold is divided into the FWD cargo compartments #1 and #2. The AFT hold is divided into the AFT cargo compartments #3, #4 and the bulk cargo compartment #5. A divider net isolates the cargo compartment #5 from the cargo compartments #3 and #4. The FWD and AFT cargo compartments of an Airbus A320 can be optionally equipped with equivalent Semi-automatic Cargo Loading Systems (CLS). Basically they have tie-down/attachment points for the nets and straps which keep the bulk cargo in place. Cargo which is to be loaded may be in containers, on pallets or loaded in bulk. Containers and pallets can be loaded in the FWD and AFT cargo holds only. Two hydraulically operated cargo hold doors which open outward are installed on the lower right-hand side of the aircraft. The bulk door, installed on the lower right-hand

There are three cargo compartments:
 • A forward cargo compartment,
 • An aft cargo compartment and,
 • A bulk cargo compartment.
The size of the fuselage accommodates standard containers.

FIGURE A1.1   Cargo holds of an Airbus A320.

side of the aircraft, is manual and opens inward. The bulk cargo compartment has tie-down/attachment points for the door nets and for the nets and straps which keep the bulk cargo in place (Figure A1.1).

BULK CARGO COMPARTMENT

The bulk cargo compartment can be loaded with loose baggage, cargo, mail, etc. A divider net with a screen separates the bulk cargo compartment from the AFT cargo hold.

Here's a good view of the lower-deck cargo hold of an Airbus A320-200 equipped with a Semi-automatic Cargo Loading System (CLS: an optional system that is electrically powered) (Figure A1.2).

The Semi-automatic Cargo Loading System (CLS) in an Airbus A320 consists of the following:

- End Stops
- Power Drive Units
- Rollers
- Ball Mats
- Control Panels
- XZ-Single Latches
- YZ Guides
- Door Sill Latches
- Entrance Guides
- Control Boxes (Figure A1.3)

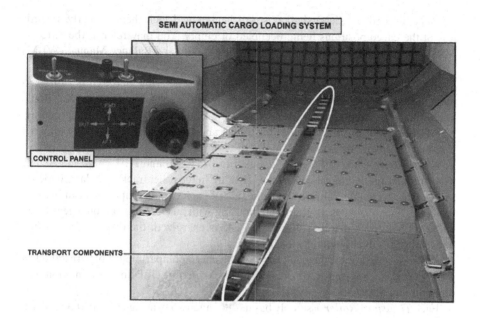

FIGURE A1.2   AFT cargo hold of an Airbus A320.

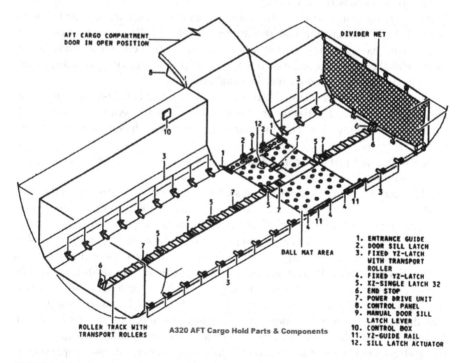

FIGURE A1.3   AFT cargo hold parts and components of an Airbus A320.

Loaders (ground handlers) shall raise all latches and lock them properly, regardless of the cargo positions being occupied or empty. You may refer to the Aircraft Maintenance Manual (AMM) and the aircraft Weight and Balance Manual (WBM) for more information.

End Stops are in the roller tracks installed on the centerline of the cargo compartment floor in the FWD and AFT cargo holds. They have the shape of fixed XZ-single latches and hold the Unit Load Devices (or ULDs that are utilized for grouping, transferring, and restraining cargo for transit and may consist of a pallet with a net or it may be a container) in the X and Z directions.

Power Drive Units (PDU) are installed in the FWD and AFT cargo holds. The PDU for longitudinal movement of the ULD are in the roller tracks. For lateral movement there is a single PDU on the right side of each cargo compartment in the ball mat area. Each PDU has an aluminum housing with an electrical motor, a gear train, and a rubber-covered roller. When the PDU is energized, the drive roller is lifted until it touches the bottom of the ULD.

Roller Tracks are on the centerline of the cargo compartment floor. Installed in the roller tracks are transport rollers. The tracks permit ULD to move in a longitudinal direction.

Each *Transport Roller* assembly has a roller and two bearings installed on a shaft. There are a washer and a cotter pin on each end of the shaft.

Ball Mats are installed across the full width of the cargo compartment floor with ball strips installed between the door sill latches. The ball mats and ball strips make it possible to move a ULD in both longitudinal and lateral directions. The ball mats and ball strips have an aluminum structure and hold the ball unit assemblies.

Ball Unit has a ball installed in a bearing shell with a circular housing. A top cover holds the ball and bearing shell in the housing also functioning as a dirt shield. The top cover has two spring struts for installation.

Linings and Floor Panels in the FWD and AFT lower holds prevent damage to the aircraft structure. The linings are made of flame-resistant synthetic material and permit fast decompression. The floor panels are of sandwich construction and have non-slip surfaces.

Rapid Decompression Panels (blow-in and blow-out) are part of the cargo compartment linings.

The rapid decompression panels are sealed to the ceiling panels and to the sidewall lining with adhesive tape. Each lower hold has a lighting system with fluorescent lamps installed in the center of the cargo compartment ceiling. Toggle switches installed at the doors of the cargo compartments control the lighting system. The loading area lights are spotlights installed in the ceiling panels at the FWD and AFT cargo hold doors. These lights permit cargo operators to read labels on loading equipment near a cargo hold door. The toggle switch for the loading area lights is installed at the FWD and AFT door operation panel. The FWD and AFT cargo holds have drainage systems which operate in an equivalent manner. Each drainage system has filters, filter holders, drain funnels, flexible hoses, and pipes. The drainage system collects and lets rainwater and spilt liquids flow out of the cargo compartments. In each cargo hold there is also a Cargo Smoke Detection System located in specific

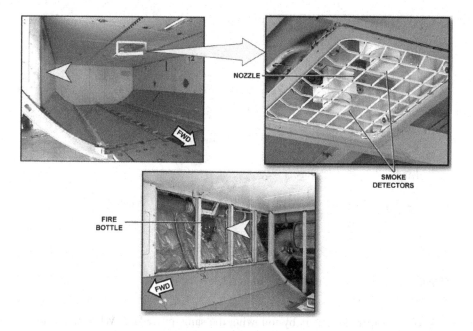

FIGURE A1.4   Cargo hold fire extinguishing system of an Airbus A320.

cavities in the cargo compartment ceiling panels: one cavity in the FWD cargo hold, two cavities in the AFT cargo hold. FWD and AFT cargo holds are also protected by a fire extinguishing system. When DISCH (discharge) pushbutton is pushed, the bottle is ignited and fire extinguishing agent is discharged (Figure A1.4).

CARGO LOADING

Here's some procedures to follow for the Airbus A320 AFT cargo hold un/loading processes. For loading:

- Switch on the compartment lights.
- Lower the two manual sill latches.
- Lowers the XZ-latches.
- Switch on the power on the control panel.

The Semi-automatic Cargo Loading System (CLS) is now ready to operate. ULDs (Unit Load Devices) lower the overridable sill latches, and when the container is over the lateral Power Drive Units (PDU), the joystick is moved to the IN position. At the end of the lateral travel, move the joystick to the longitudinal position. The longitudinal Power Drive Units are now moving the ULD. When the container is in its assigned position, lift the corresponding XZ-latch. It locks the container and de-energizes the PDU underneath it.

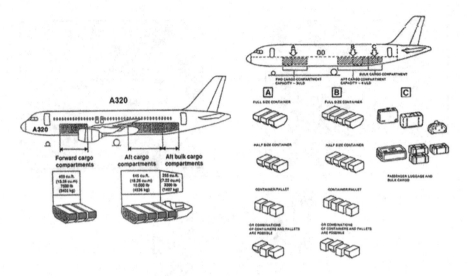

FIGURE A1.5   Cargo capacity in an Airbus A320.

Load the other containers by following the same procedure. When the last container is loaded, lift up the two manual sill latches and then check that the two overridable sill latches are up. Cargo loading operation is now completed. The power can be switched OFF (Figure A1.5).

CARGO UNLOADING

To unload, lower the manual sill latches, then switch the power ON. Lower and hold down the sill latch switch. Now the overridable sill latches are lowering. Shift the joystick to the OUT position and check the first ULD coming out. When the cargo hold is empty, switch the power OFF and lift up the manual sill latches.

MANUAL LOADING

For the manual loading of cargo, the compartment preparation is the same as with the Semi-automatic Cargo Loading System (CLS). However, ULDs need to be pushed manually instead of deploying the PDU.

AIRBUS A320 LOWER-DECK CARGO CAPACITY

The European Aviation Safety Agency (EASA), responsible for civil aviation safety and regulations, has published Type-Certificate Data Sheet for Airbus A318 – A319 – A320 – A321 [4] (the Airbus A320 Family). According to this document, the lower-deck cargo holds of an Airbus A320-200 can carry a max load capacity of 9,435 kg (20,800 lb) (Table A.1).

TABLE A.1
Cargo Holds Capacity in an Airbus
A320 (Baggage/Cargo Compartment)

Cargo Compartment	Maximum Load (kg)
Forward	3,402
Aft	4,536
Rear (bulk)	1,497

According to this table, the max capacity for each of the seven positions in the lower-deck would be **1,134 kg** ($7 \times 1{,}134 + 1{,}497 = 9{,}435$ kg). But it has to be borne in mind that there are two types of load limitations in aircraft cargo operations. According to the International Air Transport Association (IATA) Unit Load Devices Regulations [1], these load limitations are: Running (Linear) and Area Load.

RUNNING (LINEAR) LOAD LIMITATION

In order not to exceed the capability of the certified aircraft cargo compartment's floor beams and frames, the mass loaded onto any given length of the ULD base, measured parallel to the aircraft's centerline over the whole base width, must meet a maximum linear load limitation (also designated as "running load") specified in the aircraft Weight and Balance Manual (WBM) for the position it is intended to be loaded on in the aircraft. It is usually met by complying with the ULD position's maximum allowable gross mass. In practice, the *Running (Linear) Load Limit* shall be checked dividing the weight of the piece of cargo by the given length of that piece in flight direction. For example, if it is stated that the Running (Linear) Load Limit is 625 kg/m, this means that on 1 m length of the floor in flight direction, not more than a total amount of 625 kg may be loaded, with one or several pieces of cargo and irrespective of the way the piece(s) of cargo is/are in contact with the floor within the length considered. In this case, the length to take into account is defined by the external contour of its contact points:

$$\text{Weight} \div \text{Length} \rightarrow 350 \text{ kg} \div 0.70 \text{ m} = 500 \text{ kg/m} < 625 \text{ kg/m}$$

If the resulting figure is higher than the allowed limitation, the load cannot be accepted as it is, and a spreader floor will have to be provided.

AREA LOAD LIMITATION

In order not to exceed the capability of the certified aircraft cargo compartment floor structure, the mass loaded onto any given area of significant size (more than 10–20%) of the ULD base must meet a maximum area load limitation specified in the aircraft Weight and Balance Manual for the position it is intended to be loaded on in the aircraft. The *Area Load Limitation*, expressed in kg/m^2 (lb/ft^2), is to prevent the weight of the load (expressed in kg or lb) resting upon a certain area of the compartment floor (expressed in m^2 or ft^2) from exceeding the capability of the aircraft structure (floor beams, floor posts, floor panels, and frames).

In the airframe manufactures Weight and Balance Manual, such limitation is generally referred to as "Compartment Area Load Limit," "Uniformly Distributed Floor Loading," or "Maximum Distributed Load."

According to the Airbus A320 Weight and Balance Manual [2], the local loads must be separated from each other in such a way that the floor structure load limitations as well as relevant maximum compartment loads are not exceeded. The floor structure of the cargo holds is capable of supporting via the floor panels in the flat and sloping floor areas a maximum distributed load of 732 kg/m^2 (150 lb/ft^2). Each floor panel is capable of carrying a local load of 906 kg (2,000 Ib) on 0.093 m^2 (1 ft^2) without permanent deformation.

The FWD cargo hold (compartment 1) is designed for the carriage of bulk loads with a maximum load density of 240 kg/m^3 (15 lb/ft^3). This cargo hold, as per the EASA Type Certificate Data Sheet (TCDS) (Figure A1.6), has a maximum load capacity of 3,402 kg (7,500 Ib). Here's the breakdown (Table A.2).

Also, according to the TCDS, AFT cargo hold has a maximum load capacity of 4,536 kg (10,000 Ib). Here's the breakdown (Table A.3).

As a reminder, in an Airbus A320 aircraft not equipped with the Semi-automatic Cargo Loading System (CLS) in its lower-deck, the max total cargo capacity, as stated in the EASA TCDS, will be 9,435 kg. More weight information along with a

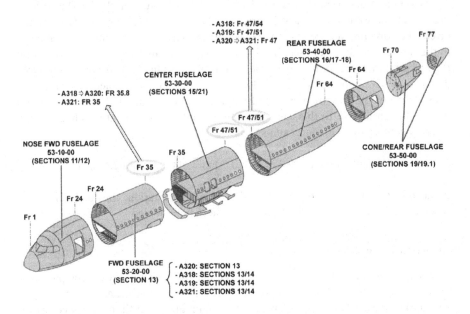

FIGURE A1.6 Airbus A320 family frames/sections.

TABLE A.2
FWD Cargo Hold Capacity in an Airbus A320

Section Designation	Extension Frame Station		Maximum Load Capacity	
	From	To	(kg)	(lb)
11	24A	28	1,045	2,303
12	28	31	1,225	2,702
Sub-Total Net Section 11 and 12	24A	31	2,270	5,005
13	31	34	1,132	2,495
Total	24A	34	3,402	7,500

TABLE A.3
AFT Cargo Hold Capacity in an Airbus A320

Section Designation	Extension Frame Station		Maximum Load Capacity	
	From	To	(kg)	(lb)
31	47	50	1,301	2,868
32	50	52A/53	1,125	2,481
Sub-Total Compartment 3	47	52A/53	2,426	5,349
41	53	56	928	2,046
42	56	59	1,182	2,605
Sub-Total Compartment 4	53	59	2,110	4,651
Sub-Total Net Section 32, 41, and 42	50	59	3,235	7,132
Total Compartment 3 and 4	47	59	4,536	10,000

weight breakdown for a typical Airbus A320-200 is shown in the Airbus A318/A319/A320/A321 Flight Crew Operating Manual (FCOM) [3] (Figure A1.7).

As we learned, the Area Load Limitation refers to the maximum load acceptable on any m² (ft²) of an aircraft floor. According to the above figures, the total max allowable capacity of 9,435 kg will be reduced to 6,491 kg with a Semi-automatic Cargo Loading System (CLS) in place. It is obvious that using the option of a CLS in the lower-deck could come with both pros and cons. Shorter turnaround times as well as the benefits of containerized cargo are among the advantages, while sacrificing more payload (consisting of the weight of passengers, including their luggage, and additional cargo) can be a significant disadvantage of utilizing such a system.

WEIGHT LIMITATIONS

Maximum taxi weight...77 400 kg (170 637 lb)
Maximum takeoff weight (brake release)..77 000 kg (169 755 lb)
Maximum landing weight..64 500 kg (142 198 lb)
Maximum zero fuel weight...60 500 kg (133 379 lb)
Minimum weight..37 230 kg (82 079 lb)
In exceptional cases (in flight turn back or diversion), an immediate landing at weight above
maximum landing weight is permitted, provided the pilot follows the overweight landing procedure.

GENERAL

The aircraft has two lower deck cargo compartments :
- Forward cargo compartment, compartment 1.
- Aft cargo compartment, subdivided into compartments 3, 4 and 5.

The main access doors to forward and aft compartments are hydraulically operated.
A bulk cargo door ◁ gives additional access to the aft cargo compartment. It is manually operated.

CARGO CAPACITY

FULL BULK

The maximum load capacity for each cargo compartment is as follows :
- **Forward**
  Compartment 1   :   3 402 kg (7 500 lb)
- **Aft**
  Compartment 3   :   2 426 kg (5 349 lb)
  Compartment 4   :   2 110 kg (4 652 lb)
  Compartment 5   :   1 497 kg (3 300 lb)

FIGURE A1.7   Load capacity per the Airbus A320 FCOM.

The following table shows the weight breakdown for an Airbus A320-211 equipped with a Semi-automatic Cargo Loading System (CLS) in the lower-deck (Table A.4).

TABLE A.4
Load Capacity in an Airbus A320 Equipped with a CLS in the Cargo Holds

Cargo Hold	Bulk – Max: 1,497 kg	AFT – Max: 2,894 kg		FWD – Max: 2,100 kg
Cargo Compartment	Comp. 5 Max: 1,497 kg	Comp. 4 Max: 1,338 kg	Comp. 3 Max: 1,556 kg	Comp. 1 Max: 2,100 kg
Max Capacity	770 kg 353 kg 374 kg	757 kg 581 kg	721 kg 835 kg	703 kg 762 kg 635 kg
Max Area Load	732 kg/m²	488 kg/m²		488 kg/m²
Usable Volume	3.04 m³ 1.38 m³ 1.46 m³	4.75 m³ 3.75 m³	4.53 m³ 5.23 m³	4.42 m³ 4.77 m³ 4.09 m³
Cargo Door Size	0.81 × 0.95 m	1.23 × 1.82 m		1.23 × 1.82 m

CONCLUSION

Weight limitations are necessary to guarantee the structural integrity of an aircraft, as well as enable the pilot to calculate aircraft performance accurately. The aircraft center of gravity (CG) is determined by summing the component weights and moments about some reference location (datum line), then dividing the total moment by the total weight. Such calculations need to be completed for each possible aircraft weight and loading configuration. An additional crew member called "loadmaster" would be responsible for placing cargo and passengers safely aboard the airplane so that the aircraft's CG stays within its allowed limits. An airliner's weight and the position of its CG are calculated before every flight. The CG moves as an aircraft burns off fuel during a flight. To calculate the weight and the CG position, the loadmaster or crew will need to know how much fuel, how many passengers, and how much cargo are onboard besides the empty weight of the aircraft. The Airbus A320 is one of the rare narrow-body aircraft whose cargo holds can be equipped with a Cargo Loading System, so the loadmaster must take this into account when doing weight and balance on this aircraft type.

REFERENCES

1. Unit Load Devices Regulations, 4th Edition by International Air Transport Association, 2015.
2. Airbus A320 Weight & Balance Manual by Airbus, 2012.
3. Airbus A318/A319/A320/A321 Flight Crew Operating Manual by Airbus, 2005.
4. Type-Certificate Data Sheet for Airbus A318 – A319 – A320 – A321 by European Aviation Safety Agency, 2020.

GLOSSARY

Center of Gravity (CG): is where all of the weight of an object appears to be concentrated. It is the point along an aircraft's longitudinal axis at which an aircraft's weight is considered to be concentrated is called the aircraft's center of gravity. Also, it is the point around which all of the moments sum to zero.

Manufacturer's Empty Weight (MEW): is the weight of the aircraft including non-removable items (also called Dry Weight).

Maximum Brake-Release Weight (MBRW): is the maximum weight at the point where the aircraft starts its takeoff run. Maximum Landing Weight (MLW) is the weight of the aircraft at the point of touchdown on the runway. It is limited by load constraints on the landing gear and on the descent speed.

Maximum Payload Weight: is the allowable weight that can be carried by the aircraft. The useful load is the sum between the payload and the mission fuel.

Maximum Ramp Weight (MRW): is the aircraft weight before it starts taxiing. The difference between the maximum ramp weight and the maximum takeoff weight corresponds to the amount of fuel burned between leaving the air terminal and liftoff.

Maximum Takeoff Weight (MTOW): is the maximum aircraft weight at liftoff, e.g. when the front landing gear detaches from the ground. It includes the payload and the fuel.

The difference between MTOW and MLW increases with aircraft size.

Maximum Taxi Weight (MTW): is the certified aircraft weight for taxiing on the runway.

Maximum Zero-Fuel Weight (MZFW): is the (loaded) aircraft weight on the ground without usable fuel.

Operating Empty Weight (OEW): is the aircraft's weight from the manufacturer, plus a number of additional removable items due to operational purposes.

Appendix – Tech Writing, Sample 2

Product Support

Aircraft Ground Handling and Airbus A320 Family

Safety is paramount in aviation. According to the ICAO Safety Management Manual, safety is "the state in which the possibility of harm to persons or of property damage is reduced to, and maintained at or below, an acceptable level through a continuing process of hazard identification and safety risk management." Evidence shows that human reliability is seriously impaired in high-pressure situations, such as airports and MRO (maintenance, repair, and overhaul) facilities, with serious time constraints. It is also important to know that there is always more than one factor involved in the humans vs. machine chain of events that leads to catastrophe. We live in a world where machines are being incrementally shaped to human anthropometrical requirements in order to ease the application of various instruments.

Performing any kind of operation in an MRO or on the airside is inherently risky, and achieving 100% safety would be wishful thinking. Risks are disruptions resulting from the unpredictability of the future caused by accidental derogation possibilities of planned targets. Hazards need to be identified and mitigated; near-misses reported and corrected proactively. Assumptions shall be prohibited because in the aerospace industry you can find documented procedures for every single activity.

Here's a good example of what you find in a typical cargo aircraft technical manual:

> Caution: Do not operate the door in winds more than 40 knots. Do not let the door stay open in winds more than 65 knots. Strong winds can cause damage to the structure of the airplane.

And some safety precautions prescribed by IATA (International Air Transport Association):

> The rubber bumpers on a loader must NEVER make contact with the aircraft. The minimum distance to be maintained at all times is 1 in/2.5 cm from the fuselage.

Or this one:

> Do not drive GSE with lifting devices in the raised position, except for final positioning of the GSE onto the aircraft. Do not drive GSE faster than walking speed.

HUMAN ERROR

Being an undesirable component of human performance, errors result from physiological and psychological limitations of humans. Eighty percent of operation errors involve human factors. Human error is suspended somewhere between the human and the engineered interfaces, and therefore new technology does not necessarily *remove* human error; it *changes* it. Since human failure can be immediate or delayed, the human element – aircraft maintenance and operations personnel, for example – must be carefully studied and skillfully managed. Active failures can have immediate consequences and are usually made by frontline people such as mechanics or ground handlers (e.g. flipping the wrong switch, getting their vehicle too close to the aircraft, or ignoring a warning tone or a safety sign). Latent failures, on the other hand, have no immediate consequences (e.g. a cargo door not being properly locked and latched after the un/loading operations).

GROUND HANDLING

Boeing has a famous definition of an aircraft:

> An aircraft is nothing but thousands of components and multiple engines, fitted to an airframe, flying in unison; which need to be maintained using Ground Support Equipment so that the aircraft can fly again and again till it is retired or in some unfortunate circumstances rendered incapable of flying.

(Figure A2.1)

FIGURE A2.1   An Airbus A320.

ICAO Annex 6 defines *ground handling* as "services necessary for an aircraft's arrival at, and departure from, an airport, other than air traffic services." Ground-handling activities at airports matter significantly to airlines because they are the most important customers of any airport. These services have an impact on both an airline's cost and the quality of services provided for passengers and freight shippers. Ground-handling services cover passenger handling, ramp handling, baggage handling, freight and mail handling, fuel handling, and aircraft maintenance and services. It is a significant fact that the process of ground handling consists of a series of highly specialized activities. In order to carry them out, highly skilled personnel, as well as suitable technical equipment, are required. Ground handling, including cargo operations services, are sometimes offered by the airport operators, although at most airports they are provided by airlines or handling agents.

Shipments vary. Time- and temperature-sensitive cargo such as HAZMAT, human remains, live animals, and perishables are kept in storage infrastructure which is part of the operator's facilities made available for the purpose of safe and proper handling of such shipments. *Unit Load Devices* (ULD) are utilized for grouping, transferring, and restraining cargo for transit. A ULD may consist of a pallet with a net or it may be a container.

Occasional ground-handling problems are very common on busy and congested airports. For domestic low-cost service by a Boeing 737 or Airbus 320, the turn-around time can be as short as 15 to 20 minutes. When performing full service for a Boeing 747 on an international route, the turnaround time can be between 1.5 and 2 hours. Some airlines have their own ground-handling agents that carry out both airside and landside services at their base or hub airports. Some airlines outsource their ground-handling services to independent service agents or to different airlines for cost-cutting purposes, especially at non-hub airports (also known as *out-stations* in the industry).

GROUND SUPPORT EQUIPMENT (GSE)

Cargo handling companies will have to have various items of *ground support equipment* (GSE), in some cases depending on the aircraft types to be supported. For example, around 21 different pieces of ground support equipment might be employed for a standard Airbus A380 turnaround or for other wide-body aircraft. Proper pre-positioning of GSE around the aircraft parking position will help to minimize the aircraft turnaround time and the potential for aircraft/GSE conflicts. Some of the well-known GSE available for passenger aircraft operations include: catering vehicle, transporter, fuel truck, passenger bridge or stairs, pushback tug, potable water vehicle, ground power unit, lavatory service truck, de-icing unit, and air conditioning. More GSE deployed for an aircraft is as follows (Figure A2.2):

Belt Loader and ULD Loader – Belt loaders are vehicles with movable belts for unloading and loading of loose cargo (and baggage) off and on to an aircraft.

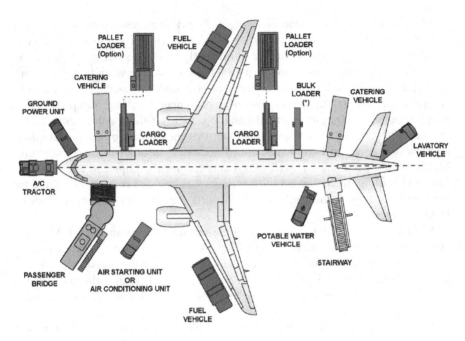

FIGURE A2.2 Typical GSE (Ground Support Equipment) found at an airport.

> **Baggage Carts and ULD Dollies** – The trolley or dolly is used to transport ULDs (Unit Load Devices, such as containers and pallets) between the aircraft and the cargo terminal. These have in-built rollers or balls to make it easier to move the loads around.
>
> **ULD Tie-Down Straps and Nets** – For cargo shipments, nets or tie-down straps need to be tight for the load to be secure.
>
> **Forklift** – Refers to the special truck that is used to lift and transport loads by means of twin tines, usually mounted on the front of the vehicle.
>
> **Loaders (main deck and lower deck)** – The loader is a platform that can be raised and lowered to enable the loading and unloading of cargo ULDs in and from aircraft. There are two platforms in a main-deck loader: one platform, usually called "the bridge," provides an interface with the sill of the aircraft cargo door; the second platform cycles up and down during the un/loading process in order to deliver cargo to the bridge.

Door Sill Rollers provide a rolling surface at the sill for containers moving in and out of the cargo compartment. The rollers also provide assistance when ground-handling equipment does not properly align with the cargo doorsill.

ANGLE OF ATTACK

Among the sensors mounted on the aircraft fuselage is the Angle of Attack Sensor (AoA) or ALPHA Probe, which calculates the angle between the direction in which

FIGURE A2.3   An AoA vane.

a wing (airfoil) is pointed and the direction of the air flowing over it – referred to as "alpha" in some technical manuals (Figure A2.3).

The AoA sensor vane aligns itself with the airstream. The discovery of the influence of the "boundary layer" – a very thin layer of air adjacent to the surface of the aircraft – was one of the most important advances in the study of aerodynamics. In commercial designs by Boeing (from the B727 to B787 Dramliner), you can find two angles of attack sensors mounted on the forward fuselage. The National Transportation Safety Board (NTSB) once recommended visual indication of AoA in the cockpit, on the Primary Flight Display (PFD). This new indicator is offered as an option on the Boeing 737-600-700-800-900, Boeing 767-400, and Boeing 777 (Figure A2.4).

Location of the AoA sensor(s) would be the manufacturers' choice. However, the air that touches these measuring devices should be free of any wake and must represent the pure flow path. In practice, the boundary layers around most surfaces of an aircraft are largely turbulent, therefore these significant measuring devices need to be in a location that is unaffected by turbulent air – a location that is exposed to as clean an airflow as possible. Aircraft nose receives the cleanest air, becoming more turbulent as you go toward the aft. When the aircraft has one engine running, the angle of attack sensors, together with other sensors such as the static ports, pitot tubes, and total air temperature probes, is electrically heated to prevent ice formation.

AIRBUS A320 FAMILY

Since the A320's entry into service in 1988, the Airbus fly-by-wire aircraft fleet, from the shorter A318 to the larger A380, has become a diverse and rapidly growing

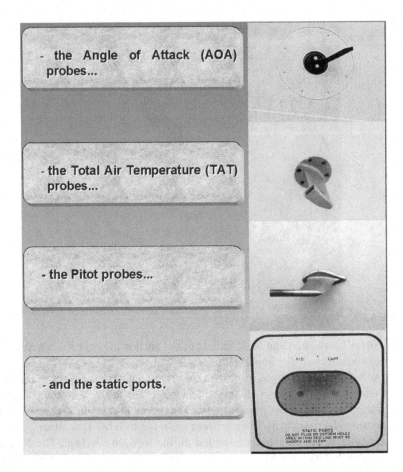

FIGURE A2.4   Aircraft sensors.

family of civil aircraft. In the A320 series, similar to its predecessors, Airbus has continued to favor three AoA sensors with only one minor exception. The manufacturer has placed two of these sensors (pilot's, number 1, and the standby, number 3) on the left-hand side of the forward fuselage, while the third sensor (co-pilot's) is placed on the right-hand side on the forward fuselage, right above the forward cargo hold. Airbus A300 and Airbus A310 are equipped with three AoA sensors, two located on the right-hand side and one on the left-hand side of the fuselage centerline (exactly similar to the A330 and the A340). In more modern generation of Airbus products such as the A380, multi-function probes are being utilized (integrating total pressure, TAT, and AoA sensors) (Figures A2.5 and A2.6).

PASSENGER-TO-FREIGHTER (PTOF) CONVERSION

Freighter aircraft are produced in different configurations. A total of 240 wide-body passenger aircraft were converted to freighters between 2004 and 2008. There

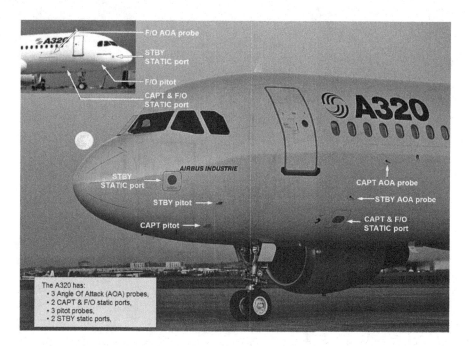

FIGURE A2.5   Airbus A320 sensors.

are over 4,000 passenger versions of the Airbus A320 in existence so there should be no shortage of this high-demand narrow-body jet to be converted. It is obvious that a Passenger-to-Freighter (PtoF) conversion involves both removing (passenger-related) and adding (freighter-related) equipment in a way that fully conforms to the Regulatory/Authority (the FAA and EASA) requirements.

It is important to know that in the aviation industry, any person who alters a product by introducing a major change, not sufficient to require a new application for a Type Certificate, shall apply to the Authority for an STC (Supplemental Type Certificate). Among the added items in a PtoF conversion is a sufficiently large cargo door that accommodates loading and unloading of cargo in the freighter aircraft. Such a door will need a frame shell for the sake of structural reinforcement of the fuselage. With the cargo door in the forward fuselage in the Airbus A320, AoA Sensor #1, located in the cutting section, will have to be relocated on the door and AoA sensor #3 will be placed on the frame shell in an A320 Freighter. Ground and flight tests prove themselves absolutely essential as a major step in this process.

GROUND HANDLING/OPERATIONS ON AN ALTERED DESIGN

Errors occur in the presence of time pressure in quick aircraft turnaround times. Working with the iron birds is fun, but by nature stressful. Each aircraft has specific areas on its structure which, by location, are particularly susceptible to damage by ground support equipment and therefore, in order to avoid damage, special

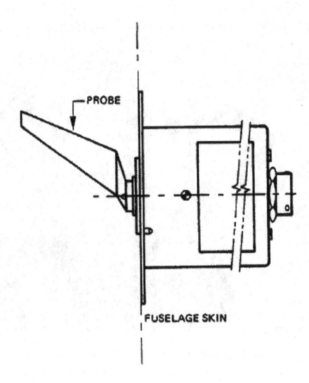

FIGURE A2.6   AoA probe.

consideration must be received. These areas include wings, flap track fairings, nose gear, engine cowlings, doors, wing-to-body fairings, sensors, antennae, and drain masts.

As a rule of thumb, when operating loaders, the bridge needs to be moved slowly until it touches the aircraft – *avoiding any aircraft sensors*. IATA Ground Operations Manual recommends moving GSE slowly toward the aircraft, avoiding any aircraft sensors, until either the protective bumpers touch the aircraft or the equipment's proximity sensors stop the movement (Figure A2.7).

ULD loaders must NEVER make contact with the aircraft and they shall be positioned **no closer than 2 in/5 cm** or until the proximity sensors (if equipped) stop the movement.

In the A320, AoA Sensor #1 is 4.30 m (169.30") and AoA Sensor #3 is 3.420 m (134.64") high from the ground (nose-wheel). The length of the AoA Sensor vane is almost 75 mm (2.95"), bigger than the 2" mentioned by IATA (above).

Since the A320 Freighter is going to have unique and unprecedented characteristics, ground-handling/operations policies and procedures as well as the training materials for all personnel would require revisions/updates to fully capture and incorporate the new safety precautions. In addition, these AoA Sensors, mainly #3, can be protected by means of some hard protective materials or covers (for the

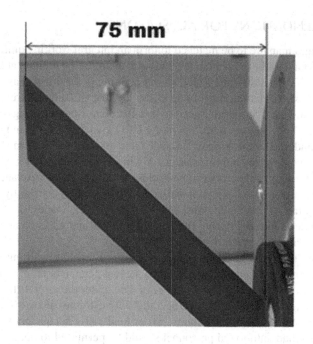

FIGURE A2.7   AoA vane length.

vane) during ground operations. Also, sensors can be clearly marked (visibility marking).

Adequate training accompanied by oversight and risk awareness would be highly recommended as part of the ground operations procedures. In this case:

> UTMOST CARE SHALL BE TAKEN WHEN OPERATING AIRBUS A320 CARGO AIRCRAFT. OUT OF THE 3 AOA SENSORS, 2 ARE LOCATED ON THE LEFT HAND SIDE OF THE FORWARD FUSELAGE: 1 IS LOCATED ON THE MAIN DECK CARGO DOOR AND 1 IS LOCATED RIGHT UNDERNEATH THE MAIN DECK CARGO DOOR IN THE VICINITY OF LOADER OPERATIONS. HITTING/SLAMMING INTO THESE SENSORS SHALL RESULT IN SEVERE CONSEQUENCES AND AOG WILL FOLLOW! THE RULES OF "RAISE, THEN APPROACH" FOR THE LOADER OPERATORS AS WELL AS "SEPARATE, THEN LOWER" MUST BE CLOSELY OBSERVED AND CAREFULLY FOLLOWED DURING UN/LOADING OPERATIONS.

According to the FAA, *Aircraft on Ground* (AOG) is a term in aviation maintenance indicating that a problem is serious enough to prevent an aircraft from flying. AOG applies to any aviation materials or spare parts that are needed immediately for an aircraft to return to service. The longer an aircraft stays on the ground, the greater the loss for the operator. Obviously anything that could reduce the ground time the operators would invest in.

RECOMMENDATIONS FOR ALL SEASONS

If machine and man are to be matched to form an integral working unit, close attention must be paid to the area of contact between them. Being vigilant at all times, we should learn to expect the unexpected when operating in sensitive areas like an MRO or an airport where maintaining an adequate level of situation awareness is highly and strongly advised. Strict rules need to be adhered to when working in a hangar or on the airside. For instance, when operating ULD loaders, IATA mandates: never to drive them underneath the wing or fuselage of an aircraft; move them slowly toward the aircraft, avoiding any aircraft sensors or wing canoe fairings; if visibility is poor or the aircraft requires the loader to be in close proximity to the fuselage or the wing trailing edge, then a guidance marshaller must be present; constantly monitor the parts of the aircraft that could come into contact with the loader; the loader bridge height shall be monitored during the loading process and adjusted as necessary to maintain a correct alignment with the cargo hold floor.

According to IATA, GSE must not be driven faster than walking speed. They need to be maneuvered carefully in order to prevent personnel injury and/or aircraft damage. And they must NOT be driven or parked under the aircraft fuselage and/or wing. As a general rule, GSE must be operated with extreme care to avoid any hazard to personnel and/or any damage to aircraft or load. Only adequately trained, qualified, certified, and authorized personnel should be permitted to operate equipment. The movement of carts/dollies by hand-operated equipment is very simple; nonetheless many injuries have resulted, and additional care must be taken.

IATA also advises operators to apply parking brakes and place the gear selector in the PARK or NEUTRAL position on all GSE when parked or positioned. When positioning GSE, make sure that clearance is kept between all GSE and the aircraft to allow vertical movement of the aircraft during the entire ground-handling process preventing contact between the aircraft and equipment. Do not drive GSE with lifting devices in the raised position, except for final positioning of the GSE onto the aircraft.

Also, among the best recommendations is providing sensor(s) under the leading edge of the bridge platform so that when contact is made with the aircraft fuselage, the forward movement of the bridge will stop.

Voluntary reporting of the yellow lights – near-misses – always helps in the prevention of incidents/accidents. To support a reporting culture, the organization must cultivate the willingness of its members to contribute to the organization's understanding of its operation. Since some of the most valuable reports involve self-disclosure of mistakes, the organization must make the commitment to act in a non-punitive manner when those mistakes are not the result of careless or reckless behavior.

CONCLUSION

In a system that is designed to prevent errors, flight crew is the last line of defense. As long as airplanes are operated by humans, fatigue, distraction, and other inherent weaknesses can only be managed, never erased. The human element is the most flexible, adaptable, and valuable part of the aviation system, but it is also the most vulnerable to influences which can adversely affect performance. Machines should be designed to make the operator's tasks possible and reasonable and replicable reducing the physical and mental strain. However, evidence confirms that situational awareness, judgment, and decision-making skills can be improved through structured training.

Bibliography

Achebe, Chinua. 2013. The African Writer and The English Language. In *Colonial Discourse and Post-Colonial Theory, A Reader*, ed. Patrick Williams, and Laura Chrisman, 428–434. Routledge.

Applewhite Ashton, William R. Evans III and Andrew Frothingham. 2003. *And I Quote: The Definitive Collection of Quotes, Sayings, and Jokes for the Contemporary Speechmaker*. St. Martin's Press.

Aristotle. 1996. *Poetics*. Translated by Malcolm Heath. Oxford University Press.

Austin, J. L. 1962. *How to Do Things with Words*. Oxford University Press.

Barthes, Roland. 1974. *S/Z*. Translated by Richard Miller. Blackwell Publishing.

Barthes, Roland. 1977. *Image, Music, Text*. FontanaPress.

Beckett, Samuel. 1964. *How It Is*. Grove Press.

Beckett, Samuel. 1995. *The Complete Short Prose*. Grove Press.

Beckett, Samuel. 2012. *Molloy*. Faber and Faber.

Beja, Morris. 1992. *James Joyce: A Literary Life*. Macmillan.

Benjamin, Walter. 1968. *Illuminations*. Translated by Harry Zohn. Schocken Books.

Bible King James. https://www.kingjamesbibleonline.org/ (accessed November 20, 2020).

Bichsel, Peter. 1997. *Kindergeschichten*. Suhrkamp.

Borges, Jorge Luis. 1998. *Collected Fictions*. Translated by Andrew Hurley. Penguin Books.

Camus, Albert. N.d. Banquet Speech. https://www.nobelprize.org/prizes/literature/1957/camus/speech/ (accessed November 22, 2020).

Carroll, Lewis. 2009. *Alice's Adventures in Wonderland*. Oxford University Press.

Chomsky, Noam. 2002. *Syntactic Structures*. Mouton de Gruyter.

Christensen, Inger. 2018. *The Condition of Secrecy*. Translated by Susanna Nied. New Directions Books.

Cummings, E.E. 1991. *Complete Poems*. W. W. Norton, & Company Ltd.

D'haen, Theo and Hans Bertens. 2011. *Shift Linguals Cut-Up Narratives from William S. Burroughs to the Present*. Rodopi.

de Saint-Exupéry, Antoine. 2020. *Little Prince*. Translated by Ros Schwarz. Pan Macmillan.

Derrida, Jacques. 1977. *Living on: Borderlines*. Translated by James Hulbert. Seabury Press.

Derrida, Jacques. 2016. *Of Grammatology*. Translated by Gayatri Chakravorty Spivak. Johns Hopkins University Press.

Eco, Umberto. 1986. *The Name of the Rose*. Translated by William Weaver. Warner Books.

Ellmann, Richard. 1982. *James Joyce*. Oxford University Press.

Federal Aviation Administration (FAA). n.d. FAA Writing Standards (Order 1000.36). https://www.faa.gov/documentlibrary/media/order/branding_writing/order1000_36.pdf (accessed November 19, 2020).

Geller, E. Scott. 2001. *The Psychology of Safety Handbook*. Lewis Publishers.

Hardy, Thomas. 1998. *The Mayor of Casterbridge*. Oxford University Press.

Harrington, H. J.. 1991. *Business Process Improvement*. McGraw-Hill, Inc.

Heidegger, Martin. 1976. *Basic Writings*. HarperSanFrancisco.

Heidegger, Martin. 1996. *Being and Time*. Translated by Joan Stambaugh. State University of New York Press.

Hemingway, Ernest. 1929. *A Farewell to Arms*. Scribner.

James, Henry. 1987. *The Spoils of Poynton*. Penguin Books.

Joyce, James. 1907. *Dubliners*. Oxford University Press.

Joyce, James. 1939. *Finnegans Wake*. Penguin Classics.

Kafka, Franz. 1971. *The Complete Stories*. Translated by Nahum I. Glatzer. Warner Books.

Khalsa, Dharma Singh. 1999. *Brain Longevity: The Breakthrough Medical Program that Improves Your Mind and Memory*. Warner Books.

Logsdon, John M. 2010. *John F. Kennedy and the Race to the Moon*. Palgrave Macmillan.

Magee, Bryan. 1977. *Facing Death*. William Kimber, & Co Limited.

Manning Rob and William L. Simon. 2010. *Mars Rover Curiosity, An Inside Account from curiosity's Chief Engineer*. Smithsonian Books.

Morrison, Toni. Nobel Lecture. https://www.nobelprize.org/prizes/literature/1993/morrison/lecture/ (accessed November 20, 2020).

Nabokov, Vladimir. 1980. *Lectures on Literature*. Harvest Books.

Naval Air Systems Command Technical Publications Library Management Program (NAVAIR 00-25-100) https://pdf4pro.com/view/navair-00-25-100-navy-bmr-508de.html (accessed November 22, 2020).

Nietzsche, Friedrich. 1985. *Thus Spoke Zarathustra*. Translated by Walter Kaufmann. Penguin Books.

Nietzsche, Friedrich. 2003. *Writings from the Late Notebooks*. Cambridge University Press.

Nietzsche, Friedrich. 2010. *On Truth and Untruth, Selected Writings*. Translated by Taylor Carman. HarperCollins.

Norman, Don. 2013. *The Design of Everyday Things*. Basic Books.

Orwell, George. 2000. *Essays*. Penguin Classics.

Orwell, George. 2002. *1984*. Signet Classics.

Perec, Georges. 2005. *A Void*. Translated by Gilbert Adair. Verba Mundi.

Pinker, Steven. 2015. *The Sense of Style*. Penguin Books.

Pinter, Harold. 1988. *Mountain Language*. Samuel French LTD.

Pirandello, Luigi. 2019. *Six Plays*. Translated by Felicity Firth, Robert Rietti, John Wardle, Donald Watson and Carlo Ardito. Calder.

Poe, Edgar Allan. 2006. *The Portable Edgar Allan Poe*. Penguin Books.

Propp, Vladimir. 1968. *Morphology of the Folktale*. Translated by Laurence Scott. University of Texas Press.

Rivkin Julie and Michael Ryan, ed. 2004. *Literary Theory: An Anthology*. Blackwell Publishing.

Rumi, Maulana Jalalu-'D-Din Muhammad. 2008. *The Mansavi I Ma'navi of Rumi*. Translated by E.H. Whinfield. Forgotten Books.

Schopenhauer, Arthur. 2004. *Essays and Aphorisms*. Translated by R. J. Hollingdale. Penguin Books Ltd.

Schopenhauer, Arthur. 2013. *Studies in Pessimism*. Translated by T. Bailey Saunders. The Pennsylvania State University.

Schwab, Gustav. n.d. The Horseman and Lake of Constance. https://www.bartleby.com/270/7/35.html (accessed November 22, 2020).

Smith, Larry. 2008. *Not Quite What I Was Planning: Six-Word Memoirs by Writers Famous and Obscure*. HarperCollin.

Stevens, Wallace. 1971. *The Collected Poems*. Alfred A. Knopf, Inc.

Todorov, Tzvetan. 1977. *The Poetics of Prose*. Translated by Richard Howard. Blackwell.

Twain, Mark. 2005. *The Complete Short Stories of Mark Twain*. Bantam Classic.

Wilde, Oscar. 2010. *The Decay of Lying and Other Essays*. Penguin Classics.

Wittgenstein, Ludwig. 2001. *Tractatus Logico-Philosop*. Translated by D. F. Pears and B. F. McGuinness. Routledge Classics.

Woolf, Virginia. 1985. *Moments of Being*. Harcourt Brace Jovanovich, Publishers.

Zhuangzi. 2003. *Basic Writings*. Translated by Burton Watson. Columbia University Press.

Zilcosky, John. 2003. *Kafka's Travels*. Palgrave Macmillan.

Index

Printed in the United States
by Baker & Taylor Publisher Services